RIVER PROFILES

RIVER PROFILES

The People Restoring
Our Waterways

PETE HILL

Columbia University Press
New York

Columbia University Press
Publishers Since 1893
New York   Chichester, West Sussex
cup.columbia.edu

Library of Congress Cataloging-in-Publication Data
Names: Hill, Pete (Watershed consultant), author.
Title: River profiles : the people restoring our waterways / Pete Hill.
Description: New York : Columbia University Press, [2023] | Includes
    bibliographical references and index.
Identifiers: LCCN 2023047143 | ISBN 9780231207645 (hardback) | ISBN
    9780231207652 (trade paperback) | ISBN 9780231557061 (ebook)
Subjects: LCSH: Stream restoration—United States—Biography. | Stream
    Ecology—United States.
Classification: LCC QH97 .H55 2023 | DDC 333.91/62160922—dc23/eng/20231212
LC record available at https://lccn.loc.gov/2023047143

Printed and bound by CPI Group (UK) Ltd, Croydon, CR0 4YY

Cover design: Julia Kushnirsky
Cover images: Top: Shutterstock. Bottom: Pete Hill.

CONTENTS

Acknowledgments   *vii*

Introduction   1

1   Fumbling for Bankfull: Dave Rosgen and the Strong Currents of a
Stream-Restoration Methodology   12

2   The Bog Architect: Reimagining Streams and Stormwater on the
Coastal Plain of the Chesapeake   38

3   Legacy Sediment: Dorothy Merritts and Robert Walter Dig
Back in Time in Lancaster County, Pennsylvania   67

4   The Human Beaver: Mega- to Micro-Engineering Solutions in
Greater Cincinnati   89

5   Beaver Wranglers: Facilitating Functional River Restoration in
Western Washington   106

6   Wisconsin Trout: Restoring Driftless Area Streams and Mitigating for
Effects of Climate Change   130

7   River Cane Dreams: A Plant That Restores Connections   160

8   Naturalized Channels in Milwaukee: Removing Concrete and
Lowering Floodwaters   181

9   Community Bonds: Organization and Collaboration in
West Atlanta   198

10   South River Action Hero: Environmental Justice and Activism in
     Suburban Atlanta   214

11   Dam Removal and Complicated Histories: Unfinished Business on the
     Elwha River in Washington State   227

     Conclusion: Returning to Watts Branch   253

     Epilogue: New Currents of the Snake River   261

     Notes   271

     Index   277

ACKNOWLEDGMENTS

Many people have assisted with this project in countless ways. First and foremost, the people who have shared their work and stories with me have made this book possible. I am eternally grateful to these people (and those who are not mentioned in the text) for sharing their time with me and answering many, many follow-up questions. Meeting them in person sustained my belief in this project and spurred me to complete it. Their particular streams and projects have imprinted on me in ways that I don't yet fully understand. Like some sort of mongrel hatchery salmon, I know I will return at some point to check in with them and their rivers and watersheds.

Thanks to Mukund Belliappa for honest comments on many drafts, Josh Liberatore and Linda Presto for very helpful advice on indexes, and April Reese for advice on shopping my proposal. Mike Rhanis, Brian Marks, Ivan Ascher, Paul Robinson, Adrian Camacho, Joanne Kuebler, Paul Muller, and Ursula Trempska have provided advice and support at different points along the way. Thanks to Nick DiPasquale, who reviewed the proposal and provided needed encouragement to someone who was only an acquaintance. Thanks to the UU Theme Circle for providing an early opportunity to voice this dream project of mine in public. A shout-out to the FFC, somehow both incessantly present and quietly supportive in the background. Thanks to Claramae Hill for her artistic offerings. Thanks to Morgan Hewitt for demonstrating courage in writing. Thanks to my editor, Miranda Martin, who has provided deft guidance and support to a first-time author.

Finally, I'd like to thank my wife, Caroline, who has been the definition of both a captive reader and an unflagging believer in this project. Like a river, this project has been nourished by these upstream sources, the unseen liquid courage that has seen me through.

RIVER PROFILES

INTRODUCTION

As a society, we are generally comfortable with our ability to modify, reinvent, or build our personal spaces from scratch. Whether it's our backyard or an open-concept kitchen, in these private spaces we enjoy shaping our immediate environment. When it comes to shared and treasured natural spaces, however, we tend to assume that our intervention would be negative and should be avoided. We protect lands deemed worthy of saving and restrict unwanted activities. We keep to the path and give nature a wide berth, but giving nature its space doesn't necessarily protect it. Increasingly, we understand that the myriad ways we affect our planet have created more systemic problems: five-hundred-year floods have devastated major metro areas such as Houston; algal blooms have shut down drinking-water intakes in Lake Erie; and western reservoirs are being depleted to never before seen levels. We have only recently begun to fully appreciate how much effort and attention is needed to ameliorate these effects.

Today our impact on the environment is so intertwined with our daily lives that it is difficult to know what we need to change. This is particularly true for rivers and streams; because of the fundamental properties of gravity and water, rivers and streams bear the load of all of our actions, intentional and unintentional. We have a dysfunctional relationship with streams and rivers. We know we need to do more, but what exactly? On some level, we know that everyone getting a rain barrel won't solve larger, interconnected water problems. Our awareness of these effects, lying at the periphery of our vision, nags at us. Despite this unclear picture, we cannot leave our rivers and waterways alone and hope that they will recover. We need to act.

Despite the invigoration such a call to action may provide, it doesn't answer the question, *What do we need to do?* Rivers and streams are dynamic and powerful forces that have the ability to reshape the landscape on their own. They also respond to and reflect the hundreds of decisions we make related to development, transportation, agriculture, and the overall management of our lands. Like a child, they reflect the thousands of interactions, influences, and resources they experience over a lifetime. Like a child, they do not always react in the ways we desire.

A group of people hailing from a variety of specialties and professions is stepping up to address this hydra-headed problem. They have answers to what we need to do for our rivers. They have created restoration methodologies, construction techniques, and engagement strategies, and they explain how their approaches make the best use of limited dollars and can be sustained over time. They promise that their ideas are exactly what is needed to begin the process of returning our streams and rivers to the life-giving and bountiful ecosystems they once were. They are at turns confident and dogmatic, visionary and paranoid. After a steady trajectory of decline, they promise to address if not fix our collective impact if their approaches are implemented. Like Hercules, who defeated the nine-headed water snake, this group possesses mythic qualities. They are facing an epic challenge of our time. It's a story we both want and need to hear.

Although they all agree that our streams should be restored, the unanimity dries up as soon as the discussion turns to particular strategies. There is no one agreed upon approach to how this work should be done. This confusion has created a situation in which those who aspire to save our streams and rivers must preach their vision and stake their claim. Channeling the optimism of a World's Fair and the lawlessness of the wild west, this field attracts self-anointed eco-saviors who differ from the "eco-warrior" stereotype. Academics weigh in with their research. Ambitious consultants angle for choice contracts. Advocates look for leverage that can bring attention and resources. Public servants weigh these factors and attempt to steer a prudent course forward. Almost universally, these individuals have a strong desire to act and the chutzpah to think they can fix these problems.

These problem solvers argue over technical differences, but the specific motivations and attitudes that underlie their personal and professional drives lead to other questions. Some bring a desire to fully understand river systems in both ecological and historical terms and work to ensure that any restoration strategy adheres to what their research has revealed. Others are more interested in building a level of support within a community that promises to sustain a river or stream in ways not specifically addressed under the scope of a restoration contract. Some government actors seek solutions that can maximize existing pools of funding or steer existing regulatory frameworks to a more ecologically friendly outcome. Increasingly, many actors are less concerned with the specifics of the channel shape when the bulldozers leave and are more interested in setting up conditions for a healthy stream sometime in the future. This might mean filling a channel with logs, creating a floodplain, or allowing beavers to recolonize a stream.

There is a tendency to gloss over these distinctions because, in the end, everyone involved loves streams. A muddled jargon within the field extols multiple benefits. There is an underlying belief that we can address multiple needs. Everyone can have it all. Yet these approaches aren't specific spices that can be tossed into a restoration project and seamlessly melded into a dish that satisfies everyone's palate. In many cases, the approaches are in direct contradiction to one another, and the science points to divergent outcomes. And most people don't want their backyard turned into a beaver pond.

This book tells the stories of the people engaged in this rapidly evolving and endlessly interesting field that I have worked in for more than twenty years. Many of the people I profile I know well. I understand the professional turf battles and the economic pressures they face. The technical differences between their various approaches point to different problems that we have unwittingly inflicted upon our rivers and streams. Although the problems are often systemic and complex, most of the solutions proposed by these practitioners are simple to comprehend. Their approaches can best be understood by pulling on hip boots and walking the streams with these engaging and committed people. This reconnaissance is the basis of this book.

The book follows the work of people restoring rivers in eleven different locations across the continental United States. I came to some of these stories over a long period of time as I managed stream-restoration efforts in the District of Columbia. This experience of stewarding limited public funds toward the restoration of long-battered streams meant choices were difficult. I spent many hours talking to some of the people profiled in this book, arguing the merits of various approaches. I heard sales pitches and reasoned arguments. I wrestled with what approaches could lead to sustainable and lasting improvements that would benefit the residents of the District of Columbia.

This search in some ways was birthed by a nagging sense of doubt. I wanted to know if what I and my colleagues had accomplished was worth not only my time but also the countless contributions of many people. I wanted reassurance that the millions of tax dollars that were being spent were being spent wisely. I wanted to know if the projects improved the lives of people living near the streams. Fundamentally, I wanted to be reassured that the work we had undertaken was sustainable. These doubts led me to the people I feature in this book.

Although their approaches typically originate out of a specific area, they resonate beyond their origin with the promise of applicability throughout the country. The people behind these efforts are unique and highly driven and on a mission to spread their specific solutions. As an informed guide, I let them explain how their particular solutions might help us get a handle on this overwhelming challenge.

Whether you are involved in this work or not, when you have an understanding of the approaches and how they might expand across the landscape, you will have a stake in the game. Despite the expertise profiled here, the informed support of the broader public is critical to realizing the vision of these individuals. I hope that this understanding will lead you to become more engaged in this long-term effort to restore our waterways.

Ultimately, we'll need to implement multiple approaches to effectively restore our streams and rivers. This may seem obvious, but this conclusion contrasts with the strong tendency in the field to identify one right approach (with most practitioners believing their approach is the correct one). The contrast in approaches (low-cost approaches, habitat-focused approaches, keystone-species-dependent approaches) also points to the issue of how the work is implemented. The low-bid contracting approach that defines the trajectory of most projects may not be the best economic model. Some approaches promise to take advantage of time, and others require massive and costly earthmoving to address effects from centuries past; it is difficult to evaluate these different approaches based on cost alone. The presumption of certainty that undergirds most review processes strains credibility when few projects are monitored a full five years after the construction crews have packed up and left. With such an abundance of energy and ideas in this dynamic field, it is incumbent on us to find a model that outlines appropriate approaches and where they are best suited. Throughout this book, I'll help you understand the relatively unseen world of how work is prioritized, designers are selected, and projects are implemented.

My profiles begin with the most outsized personality in the field, Dave Rosgen. A former Forest Service employee without formal academic training, Rosgen left his work and began assessing streams in the Rocky Mountains. His many hours of observation led him to develop a classification system that described the changes streams go through, frequently following human impact. His system enables people, after a relatively short period of study, to label streams as certain types. No such detailed and proscriptive system for classifying streams existed before this. Using Rosgen's system, people talk about a "C4" stream, and others understand what type of stream this is. People are also able to read into past disturbances, which is critically important for rivers where records are scant or unavailable.

Rosgen quickly spread the word about his clever methodology by creating a series of week-long classes that he taught with assistance from a small number of trusted staff. His classes fit the bill for government staff and consultants who wanted guidance on their numerous stream problems. The classes were also rousing events, a blend of summer camp and a heady intellectual atmosphere. The courses quickly became prized trainings, and thousands of people flocked to his classes.

Rosgen's approach was bolstered by government staff who used his trainings as formal criteria for request for proposals (RFPs). Having paid for many failures in the past, funders wanted to be sure future projects were completed by qualified people, and Rosgen's approach seemed to be an enlightened way to secure that level of involvement. The combination of government requirements, the need for a clear methodology, and the relative understandability of the system made Rosgen a wealthy man. It also made him a target.

Academics were the first to aim their fire at this person who did an end run around their review and approbation. They painted him as a cowboy huckster who was cashing in on a system that didn't reflect reality. Many academics pointed to streams in areas untouched by human impact that had been in flux for centuries. They questioned his profit motive. They accused government reviewers of essentially forcing people to take his expensive classes in order to be considered for work.

Others saw the problems in the streams differently. Dorothy Merritts and Robert Walter, a husband-and-wife team at Franklin and Marshall College in Lancaster County, Pennsylvania, made the discovery that the form of streams in much of the mid-Atlantic was an artifact of widespread damming of the rivers by early colonizers. Through their research, they showed that many streams were essentially buried by "legacy sediment" that eroded from deforested lands in headwaters (many decades past) and settled behind low-head, millpond dams. This accumulated sediment changed streams from a vibrant, multichannel wetland system abundant with wildlife and clean water to single-stream channels that had far poorer habitat and water quality. Not only did they advocate forcefully for removing this sediment but they also launched projects in Lancaster County to do just that. Their projects removed thousands of tons of sediment and created braided river channels that looked vastly different from the channel that had existed for the past 150 years. In this restored form, they provided ecological benefits that hadn't been seen since before the colonial period. Their work shifted the restoration target chronologically to a time in the past that few academics or practitioners had considered. It was a significant contribution that has raised major questions about the extent of historical and ecological purity to which restoration should be held.

Battling a different set of problems with a fervor bordering on evangelical is a restoration contractor in Annapolis, Maryland. Keith Underwood, trained mostly on the job in landscape contracting, began working in smaller, highly eroded streams in suburban areas. These streams were primarily affected by unregulated stormwater runoff from parking lots, driveways, and roadways. The effects were alarming; in some places the bottom of the stream had dropped thirty feet from its original elevation. The displaced sediment made its way to the inlets of the Chesapeake Bay, thwarting broader restoration efforts.

Underwood saw an opportunity in this communal neglect. He proposed and implemented several projects, filling in these suburban canyons with a mix of sand and woodchips. The stated goal was to slow and treat this torrent of stormwater. His projects revealed a more ambitious goal, however, that of reestablishing rare bog habitat in unlikely suburban enclaves. In doing so, Underwood made the most of his knowledge of materials and construction methods. He used materials in ways unheard of before and created environments that promised high ecological diversity and bounty. He also worked with an eye for aesthetics—creating restoration projects that exuded a specific feeling of nature that resonated with many.

This approach, both practical and inspired, flew in the face of regulatory requirements and quickly drew the ire of regulators and other practitioners. Underwood is as polarizing as they get—denounced in some circles and praised as a visionary in others. Obsessed with white cedar bogs, Underwood realized that the newly created sandy saturated environments could act as refugia for this forest type, which had become rare in the past two hundred years. A complicated personality, Underwood feeds off this controversy as he creates disciples and enemies as lasting as his re-created ecologies.

Other approaches to stream restoration have arisen as well. In an attempt to meet the need for low-cost projects, Bob Hawley developed a system for placing logs in eroded headwater streams. Acting as a human beaver, Hawley's goal is to capture future sediment in these log jams that will enable the stream to build up to a stable level over time. His work acknowledges the temporal trajectory of streams—features that respond to changes over a long period of time. It also implicitly questions the staggeringly high costs of typical stream-restoration projects, which run from $250 to $500 per linear foot, or $1.3 million to $2.6 million per mile.

In certain situations, one species can affect entire river systems. Beavers have been a key factor in determining the form and function of rivers, and where beavers have been reintroduced streams have shown major adjustments in channel form and streamside vegetation. I highlight the efforts of the Tulalip Tribes in western Washington State, where beavers have been relocated to specific headwater streams. The transformed conditions brought on by these original ecological engineers can lessen the impact of drought, provide fire-breaks, and create habitat for macroinvertebrates that in turn support threatened salmon. The role of the Tulalip in reestablishing this keystone species points to the importance of indigenous cultural knowledge that is not often used in restoration.

I also highlight the efforts of the Milwaukee Metropolitan Sewerage District (MMSD), which has promised to remove nearly fifteen miles of concrete channel that was installed in the 1940s and 1950s. I profile the executive director, Kevin Shafer, who has been with the agency for nearly thirty years and has been a key proponent of this work. Shafer breaks the mold of a top utility bureaucrat. A relentlessly optimistic leader, he is pushing a previously uninspired sewer department toward lofty sustainability goals and national recognition. MMSD's work is unique in that the reduction of flood risk is the unifying goal of their work. This goal allows MMSD to undertake related efforts such as property acquisition, stormwater management, and upstream conservation easements. It also allows MMSD to allocate vast sums of money toward this work.

I traveled to the other side of Wisconsin to better understand the ongoing restoration of trout streams. This work, implemented by the Wisconsin Department of Natural Resources (WDNR) "strike team leader," Nate Anderson, seeks to improve habitat for the prized native brook trout while also attempting to address the worst effects predicted from climate change. With the support of his team of four and an impressive arsenal of heavy machinery, Anderson cranks out miles of stream restoration each year, particularly impressive considering the long mud season and even longer winters of northwestern Wisconsin.

Anderson's work is supported by a number of partners, first and foremost by the local chapter of Trout Unlimited. By raising funds for portions of the work as well as engaging in the physical work of razing unwanted vegetation, these homegrown chapters play a significant role in stream restoration. One volunteer in particular, Kent Johnson, has been monitoring numerous restored and control streams for more than twenty years. The data gathered not only helped WDNR improve their techniques and better quantify the results of their work but also convinced the nearby town of River Falls to adopt one of the most protective stormwater ordinances in the state.

Although the restoration of rivers frequently taps into a deep if opaque well of emotions, rarely does restoration help reestablish a cultural tradition that is central to a people. The work of Adam Griffith, an employee of the Eastern Band of Cherokee Indians, to incorporate river cane into stream-restoration projects does just this. Once covering vast swaths of river bottomlands in the south, river cane brakes now occupy only 1 percent of their former acreage. In addition to providing significant water quality and habitat benefits, river cane was used by many native tribes for everyday items as well as for highly valued cultural objects. Present-day Cherokee basket makers exemplify a connection to a resource that is both practical and profound. This different kind of relationship, flowering in a resurgent Cherokee nation, points to a model of connection with rivers that incorporates tangible use of a resource with a responsibility that spans ecological, cultural, and spiritual realms. This work raises questions of how we might treat rivers if our connections were more immediate—if those practices we valued most depended on our local rivers and streams.

The work in restoring streams in urban Atlanta requires approaches that go beyond the technical. How can one raise hope for the full restoration of an urban stream when the development of one's city has fundamentally altered it? The expansive and conflicting agendas that come with this density of human activity can make the restoration of urban streams seem out of reach. On different sides of the coastal Atlantic and the Gulf of Mexico watershed divide that runs through Atlanta, two different approaches are well underway. On the west side, a loose and intertwined group of community organizers has been raising support for a range of activities that will ultimately determine the health of the urban streams found there. This area has been plagued by devastating flooding along with massive economic and social challenges, but it is turning a corner. Parks that incorporate stormwater management are replacing dilapidated and vacant housing. Organizations are training and empowering local residents in the vernacular of watershed restoration and stormwater management. The initial results are impressive albeit still incomplete. On the east side, the leader of the South River Watershed Alliance, Jackie Echols, has been pulling whatever levers of influence she can access. By alternately badgering public officials and courting them, Echols has repeatedly raised the uncomfortable reality of environmental injustice to anyone who will listen. The neglected river that she champions has borne the brunt of the effects of urban development. Part of her time is focused on getting people out on the river. Another part is pointing out the unique challenges that her river and watershed face, challenges receiving less attention than similar ones in more affluent watersheds. This good cop/bad cop routine is beginning to yield results. She reminds us that this journey to restore our rivers frequently takes a different route depending on the complexion of the residents living there.

Dam removal is perhaps the most iconic and singular form of river restoration. The Elwha River in the Olympic Peninsula was one of the most prolific salmon fisheries in the Pacific Northwest before installation of two massive hydroelectric dams. The river supported ten distinct salmon runs, including the native Elwha salmon. These dams were installed despite the historic interdependence with this river by the local Lower Elwha Klallam Tribe. For the past twenty years, the Lower Elwha Klallam Tribe has been deeply involved in this long-term restoration project. Contrary to conventional wisdom, the removal of a dam is not the end of a river restoration effort. Their fisheries biologists are currently installing habitat structures and are actively managing salmon stocks in an effort to "rewild" the salmon stocks that have been hatchery bred for decades. This effort points to the long-term role that may be required if we want to restore not only the river channels but also the full assemblage of species that use them.

My personal connection to this particular river comes from a period during college when I frequently camped and hiked in the area. At the time, the structures appeared as an immovable testament to how we have permanently damaged our environment. Looking up at 210 feet of vertical concrete, I saw a massive stone tablet whose message was clear despite the lack of words. I felt little hope that we possessed the ability to rectify our impact. Fortunately, many people began a tireless process of planning that stretched out over twenty years. The process wound through controversy but eventually was successful. During a recent trip to the Pacific Northwest with my wife and daughter, I returned to the site and had unfamiliar feelings in a now unfamiliar landscape. Instead of slow and steady decline, I witnessed evidence of heroic and incontrovertible improvement. Rather than wearying dread, I saw justified hope. The freed river clearly attests to the fact that we possess the ability to reverse some of our greatest damages. Yet the work goes well beyond just removing two massive impediments. In addition to bringing technical and cultural knowledge, the members of the Lower Elwha Klallam Tribe bear witness to the need to right a historical wrong. The questions currently being raised in this corner of the continental United States also reveal the complicated, interconnected dynamic of salmon harvest. The experiences of those still working on this restoration effort reveal how challenging it is to fully restore a river.

Finally, I returned to Washington, D.C. to revisit a stream-restoration project I managed for more than eight years. The stream wound through a neighborhood that had seen more than its shares of challenges, and I encountered many of these as I worked to get "dirt moving." This project involved years of coordination, mind-numbing bureaucratic delays, and millions of dollars. The project was eventually implemented, but my experience raised questions of sustainability and the community effects that point to the broader question of what we are doing with these projects. The distance of ten years has brought a perspective that helped me clarify these questions.

These are the people and the stories that underlie these stream- and river-restoration efforts. Each one is as unique as the streams they steward. With an abundance of committed individuals involved in this field, it has been difficult to limit myself to eleven chapters. Necessarily, I have chosen people whose approach or background can provide a broader perspective on the issues our rivers face. This book is not intended to identify the most authoritative or accomplished individuals in stream restoration but rather to showcase people who are telling important stories.

In interviewing these restoration practitioners, I wanted to know both the specifics of their designs and the contour of their experiences. I reached out to academics, and I peppered consultants with questions. As I began to talk to others, I discovered fabulous projects championed by unique and driven people. I saw traces of god complexes, and I witnessed painstaking analysis. I saw people taking risks, some being rewarded and others facing ongoing criticism. I realized that the problems these individuals were trying to solve were shaped by local and regional biological, cultural, and historical contexts. I saw a field that drew upon numerous disciplines and one that was in flux. I wanted to know more, and I wanted to share this information.

This book is a product of that search. It's the product of a slow-boil obsession, an education produced from a network of willing teachers who have let me into their world. These stories of people engaged in the restoration of our waterways carry a strong current of hope upon which I gladly floated. The efforts I describe rekindled my passion for this work. Streams and rivers represent the ultimate litmus test for how we live on this planet. Rivers gather everything we throw at them and do their best to absorb it. In understanding the efforts made to restore them, we come face-to-face with our own behaviors and the weight of our impact on the land. If we want our liquid arteries to sustain us and the rest of the living world, we, like the people showcased here, need to step up to the challenge.

CHAPTER 1

FUMBLING FOR BANKFULL

Dave Rosgen and the Strong Currents of a
Stream-Restoration Methodology

TOOTHPICK SURVEYS AND TEACHABLE MOMENTS

The group forms a semicircle around a figure whose back is to them as he peers out onto the river below. Wearing a large cowboy hat and a gigantic belt buckle to match, he cuts a clear profile. Dave Rosgen, the cowboy river whisperer. People edge closer, silently surveying the floodplain in an attempt to see what he is seeing. He is the reason everyone is assembled on this riverbank. Everyone is ready to take in his thirty-five years of accumulated knowledge. He exudes a kind of energy, confidence, and general lack of bullshit that strike the major chords of American determinism. He turns to assess his fresh group of recruits and flashes a smile that puts everyone at ease. He is in his element. This is not his first rodeo.

"The first order of business is our channel survey," he happily announces in the same tone as if he were asking us to hop on quarter horses and pace the property boundaries. The "toothpick" survey as he calls it. Without a hint of irony he actually puts a toothpick in his mouth and hops down the bank into the active stream channel. "Come on down—you're not going to understand a river without getting your feet wet!" The group makes its way down like a herd of cattle desperate for water.

We follow Rosgen downstream and observe. He is reading the watery landscape, its manuscript composed of the water and the sediment that is moving underneath our feet. While our leader follows the current, he points out drift lines of dead vegetation, vegetative detritus pushed into the river's edge by recent floods. He points to riffle sections and picks up some of the underlying material, a coarse cobble that would be ideal for spawning. He explains channel adjustments that happened several months ago and others that happened several years ago. His vision allows him to both peer back into time and into the future. It's a superpower of ecological omniscience—part detective, part prophet.

Despite the slippery boulders, the random loose rock, and unexpected pools, we follow eagerly. People jostle for a closer position to better see what our leader is seeing. Some make observations to others, friendly small talk that also telegraphs that they've had previous exposure to the system and already get it. Rosgen's confidence is contagious. The collective mood is energized. We are approaching the source, these proverbial headwaters of knowledge and understanding. We are seeing something very few understand. It's becoming clear to us how water has shaped and is currently shaping this constantly changing river. All of us, for our own professional and personal reasons, need what we are receiving right now.

We come to a point in the stream where sand and small cobble has been pushed up into a band of nearly impenetrable alder shrubs that line the channel. Our leader turns to the group and asks, "Who knows where bankfull is?" No one in the group responds. These are people well into their careers, some responsible for multi-million-dollar budgets and large staffs. The soft gurgling of the stream belies a grad-school tension of not wanting to be wrong. "How about you?," and he points to me. An audible guffaw from another white guy perhaps ten years my senior. A snicker from a middle-aged woman enjoying the show of testosterone. I had a feeling that this might be coming. I walk out of the stream and over to a band of shrubs where fine sediment has accumulated. Trying to muster some confidence, I walk out of the river onto dry land. "Right here along these shrubs?" I ask, as I pick up fine grains of sand that fail to reflect the bright sunlight.

Rosgen smiles and shakes his head and scratches his temple. The wind whips up some dust, hailing like tumbleweed from some high plains source only known to him. "Don't look at the vegetation. It can fool you," he offers. I quickly drop the offending grit from my hand. His tone is not critical, just John Wayne confident. With the energy of an unruly bull, he pushes through the wall of shrubs next to the stream and stops at a flat spot another few feet past the edge of the vegetation. He tramples down the saddle-high shrubs so others can follow and see. The flat area is covered with a fine, light-brown, silty residue left from the last flooding event. "Right here. This is bankfull," he declares (figure 1.1). There is relief among the group; it has been revealed. The crowd follows his path through the vegetation to get closer to the spot. People examine the dirt, pick it up, and pass it around. It is a heterogeneous mix of sands and silts nearly identical to what I had picked up. Yet this dirt rings true whereas mine carried the falseness

STEPS: 1. Obtain a ROD READING for an Elevation at the "MAX DEPTH" Location.
2. Obtain a ROD READING for an Elevation at the "BANKFULL STAGE" Location.
3. Subtract the "Step 2" reading from the "Step 1" reading to obtain a "MAX DEPTH" value;
then multiply the Max. Depth value times 2 for the "2x MAX DEPTH value.
4. Subtract the "2x Max. Depth" value from the "Step 1 Rod Reading" for the
FLOOD-PRONE AREA Location Rod Reading. Move the rod upslope, online with
the cross section until a Rod Reading for the Flood-Prone Area Location is obtained.

SURVEYOR'S ROD

④

②

SURVEYOR'S LEVEL

①

FLOOD-PRONE AREA   Elevation / Width ⑤

2 x MAX DEPTH

BANKFULL STAGE ⑥

MAX DEPTH

5. Mark the Flood-Prone Area locations on each bank. Measure the DISTANCE between the two "FPA" locations.
6. Determine the DISTANCE between the two BANKFULL Stage locations.
7. Divide the "FPA" WIDTH by the "BANKFULL" WIDTH to
calculate the ENTRENCHMENT RATIO.

1.1 Diagram from *Applied River Morphology* explaining how to correctly measure important channel features in the Rosgen classification system.
Courtesy of Wildland Hydrology.

of fool's gold. The recruits stand on this consequential patch of sediment, nodding their heads and sensing its significance. They drink in our leader's certainty and assess the stream channel from this new perspective. The full importance will be revealed to all over the course of the coming week.

APPLYING THE SCIENCE AND BUILDING AN EMPIRE

Dave Rosgen and his tightly knit team of family and friends have been conducting applied river geomorphology classes for more than twenty years. His company, Wildland Hydrology, organizes these short courses throughout the United States. Depending on the level, these trainings are either one or two weeks and are typically held in beautiful wilderness settings. Accommodations are comfortable but not extravagant. As long as there is good river access and a place for a bonfire, and both are not situated in a dry county, Rosgen can do his thing.

People come with a purpose, and Rosgen meets that purpose head on. Whether they represent a government entity responsible for fixing a problem or a consultant who wants to design the solution, their needs are real. These people have river problems. They have blown-out streams, incised streams, streams with no habitat diversity, and streams with massive sediment bars. Back home their needs are immediate and pressing. Residents have called their local agencies and contacted their political representatives. Problems have been "elevated."

The training is rigorous yet applied. It's not full of academic theories, nor does it contain endless qualifiers in need of further research. It's designed to provide a system for fixing streams that can be applied by anyone with the inclination to learn it and the time to assess the stream in question. Until Rosgen, the science of fluvial geomorphology—the study of the interactions of the physical shapes of rivers, their water and sediment transport processes, and the landforms they create—hadn't delivered anything particularly useful for the people ultimately responsible for these problems.[1]

It wasn't as if people weren't studying streams or were ignorant of the problems that were showing up everywhere. Principles of fluvial geomorphology had evolved from the early work of William Morris Davis, a Harvard professor whose "geographic cycle" theory proved to be highly influential in the developing field of geomorphology. His 1889 paper, "The Rivers and Valleys of Pennsylvania," described the temporal component of river channel formation and categorized rivers into youthful, mature, and old-age stages. His theory dominated the thinking of geomorphologists in the first half of the twentieth century until later geomorphologists criticized his assumptions. Davis was known to react "violently and disdainfully" to criticism, a touchiness that seems to permeate the field to this day. His mainly qualitative theory enabled people to extrapolate from these larger erosional processes down to what was happening in a specific reach. Although Davis recognized that the type of rock and erosion played a part in how landforms developed, he claimed that the most critical factor in this process was time. Current government staff tasked with saving a roadway threatened by a migrating riverbank might not find this theory as useful or timely.

A contemporary of Davis, Grove Karl Gilbert, addressed these questions from a slightly different perspective. A geologist from the era when geologists roamed the American West on horseback, Gilbert helped found the United States Geological Survey (USGS) and was appointed its chief geologist under fellow explorer John Wesley Powell. His approach was more mechanistic and examined the particular components that contributed to landform development. His work responded to the environmental degradation of the time. His 1914 study, "The Transportation of Debris by Running Water," essentially created the field of sediment transport. The flumes and raceways he developed at the University of California, Berkeley to model stream processes were a response to the devastating hydraulic mining that took place in the Yuba River in the Sierras from 1855 to 1884. Researchers have been fine-tuning his quantitative methods ever since. Despite his outsized contributions, the focus of his work was on describing the empirical relationships between water and sediment in rivers. The thought of offering applied information that could be used to restore rivers was still eighty years away.

The field added an important quantitative approach with the work of Reds Wolman and Luna Leopold in the 1950s and 1960s. This pair collaborated extensively and provided a better understanding of how and why rivers change. Their work involved significant measurement not only of sediment sizes but of rates of transport. This work was granular in every sense of the word. Their work involved measuring angles of bank toes, determining the median particle size in the streambed, and measuring in the field the amount of sediment that was passing through a stream channel. This involved measuring sand grains and burying sediment traps in the middle of stream channels. They wanted to show that streams were affected by development and other actions in the watershed. They had seen river managers reduce complex fluvial systems to pecuniary volumes of water to be allocated to those possessing water rights. They had seen planning commissions approve projects with hundreds of acres of associated impervious surface that turned healthy streams into degraded ones within a few years. There was a mission behind their work, but their stated focus was on developing the scientific underpinnings of fluvial geomorphology. It was a time when proven science could dictate future actions, presumably for the better.

Their work also developed the concept of a bankfull discharge, which was the discharge that was the most effective at passing sediment through the stream and was responsible for its geometry. For the general public, the idea of sediment moving in streams may seem undesirable and problematic at first glance, but it is essential for streams and rivers to move sediment. The problems arise when there is too much or too little. This is when culverts clog with sandbars or stream banks calve away like icebergs, threatening homes, roadways, and bridges.

This concept of bankfull discharge contrasts with a commonly held assumption that major floods determine the channel form. Leopold and Wolman acknowledge that floods can modify streams, which can be witnessed in the aftermath of any significant flood. But floods damage channels when human activity fills in floodplains, pave over their watersheds, or otherwise alter the "boundary conditions" of a watershed. In an unaltered state, the energy of a one-hundred-year flood would dissipate when the stream overflows its banks. The flow that conveys sediment in a steady and sustainable manner is the bankfull flow that comes around every year and a half on average. By shaping this transport of sediment, this flow creates the channel form. This concept of a flow that moves sediment was coined "dynamic equilibrium," and it is fundamental to Rosgen's system. It calls for understanding what form the river should be, given its bankfull flow. Once this is known, a designer can create the channel that the river "wants to be." In most rivers and streams where restoration is being considered, the challenge is that the channel has veered far from its ideal form.

WORKING UP THE DATA

It's late in the afternoon, and the group has broken into four teams that are working intently. The team members are positioned along a long yellow tape measure that spans the channel and extends up along the bank. One person handles a survey rod that screeches loud electronic beeps that pierce through the hypnotic gurgling of the steam. At first, the auditory missiles are spaced by long pauses. Over the ensuing moments, the distance between the pauses shortens, building to a mind-numbing flatline whine. This sonic pattern adds a level of intensity and annoyance to the process. The process continues as the rod-holder moves along the channel measuring key points in the streambed. When the survey rod flatlines, the rod-man breaks the peal of sound by tilting the rod and shouts out a number: "10.45!" Standing in a shallow eddy, another team member writes down this number. "Is that top of glide?" she asks. "Yes, I think . . . wait . . . let me take this one." The process repeats itself and builds to another painful auditory crescendo.

Another team member inspects the area on the far side of the channel from the others. She notices that the stream seems to be deeper on that side than in the area where the rod-holder is. "Hey, should we shoot this elevation? It seems like the channel might be deeper over here." The rod-holder heads over to shoot the area in question, and it turns out to be deeper. The stream features they are standing on seem to be in flux at this one point in the stream. Certainty fades when the ground underfoot is uncertain. Nearby, Rosgen stands in the channel, secure as an ocean piling, and observes the discussion. Like any good teacher, he doesn't reveal the answers immediately.

After a few more survey readings, he tells us the stream is forming a new channel on this inside bend. The stream is taking its increased energy and is bypassing these pool and riffle features. He tells us to take all of the survey points so we can figure it out when we work up the data. Rosgen does his best to explain a fluvial process that we are in the middle of both chronologically and spatially.

Despite the confusion and the spine-tingling sounds from the survey rod, everyone is enjoying themselves. Cumulus clouds above slowly drift to unknown basins, but the mysteries of this particular river segment are slowly opening up. A student examining a riffle picks up some of the cobble underneath. "Is this bed material appropriate for rainbow?" someone asks. Yes, and the proper channel form in dynamic equilibrium keeps this gravel bed material in place. There is power in being able to categorize something that you previously didn't know how to categorize. Knowledge expands exponentially, along both axes of scientific understanding and biophilic love.

It's clear that the processes we are assessing affect so much beyond our small encampment on this reach: fish spawning, the populations of macroinvertebrates that sustain these fish, the export of phosphorus attached to sediment particles that will flow downstream through three states and end up in the Chesapeake Bay. This phosphorus will determine water clarity, the size of dead zones, and whether crabbers can make a living next year. Our simple survey data points extend and connect to these larger consequential issues. They also point back toward us, creating a scatterplot of our own collective stewardship of these streams. Standing in the water whose current pulls our awareness downstream, we continue our measurements, realizing that we're capturing something bigger than what lies under the yellow tape stretching across the stream.

Now a full nine hours into our field day, Rosgen asks us to try to wrap things up. My team has had some prior experience using this particular survey equipment, so we have completed our cross section, longitudinal section, and pebble count. But we haven't had a chance to plot it out and assess its significance. Tomorrow is a lecture day beginning at 8 a.m., and we'll be sharing our data with the class. It's clear we'll need to work late tonight, but no one seems to mind. "You guys can work up the data at the bar!" Rosgen offers. We're deep in West Virginia, far from home and family, with little else to do. If this is what is required to understand rivers, we're all in.

THE SYSTEM

Rosgen's classification system created a new language for talking about rivers. It provided a system that could categorize a stream type using a handful of measurements. These types were universal, meaning that a Rocky Mountain "B4" stream in Colorado functioned with the same geometry as a Piedmont "B4" stream in Virginia. This universality aided comparison but most significantly allowed river-restoration designers to come up with recommended channel geometry that had a specific reference. It provided a sense of authority in contrast to the ever-present second-guessing of a field littered with highly visible failures. It resonated with biologists who could see that the ecological target for a proposed restoration project was a healthy stream that currently existed, a sort of ecological doppelganger that possessed all of the biotic attributes they valued. It provided government funders and regulators with a standard from which to evaluate proposals; and it provided an all-important rationale for their policy and contracting decisions. It gave consultants a ready-made design template that was quickly understood by those awarding contracts and issuing permits.

Finding certainty in rivers is no small feat. Diversity in stream types is vast. An arroyo in New Mexico that dries out in summer evokes a completely different landform from a Louisiana slough whose dampness pervades into all seasons. When considering the human element, the focus evaporates as professional disciplines multiply. The range of types of knowledge related to rivers and streams spans numerous fields. Macroinvertebrate biologists turning over rocks in search of caddisfly larvae rarely interact with USGS hydrologists who monitor decades of stream flow data. Scientists examining the carbon processing at the stream's edge rarely engage with engineers who are looking at sheer stresses on these same stream banks. To develop a system for classifying streams as well as a template for restoration would require that everyone gather the same type of data and agree to follow what that data was telling them. It would require people to start reading rivers in the same manner.

No part of this system was developed in an ivory tower. Rosgen developed his system during his years of fieldwork in streams and rivers in the Rocky Mountains. He spent hundreds of hours walking streams in Colorado, first in his job with the U.S. Forest Service and then on his own. In his classes, Rosgen relates that he was out of the office and on his way to a field site before his supervisor got to work. These seemingly mythical days in pristine streams, presumedly far removed from any government forms and paperwork, were spent surveying, measuring, but most of all observing. Over time, he would devise a system that was clear, transferable, and most important applicable to the budding river-restoration field. Akin to a dichotomous key created by Linnaeus to identify plants, his system satisfied a scientific longing to categorize, to place a constantly evolving thing into a tidy box. Although Rosgen describes these early days with fondness, he wasn't satisfied with measuring his Rocky Mountain streams. As a man possessing convictions and a gift for communication, he had to share his system. This system that made so much sense to him could be taught, and he knew he had an audience.

This system is the foundation of all of Rosgen's classes. At its basic level, the classification system uses a handful of measurements of the channel that, with the appropriate equipment and survey skills, can be made in a few hours. These measurements are turned into dimensionless ratios that enable people to compare one stream in North Carolina to another in North Dakota. Furthermore, these ratios can fit into predetermined bands that allow practitioners to classify them (figure 1.2). These classifications, or stream types, are a sort of archetype of streams. Think of a lazily meandering stream in northern Minnesota, northern Michigan, or in the foothills of the Rockies in Colorado with sandy point bars and deep pools across from the bars. That's a "C4" channel type. Imagine a relatively straight boulder and cobble-strewn stream, something out of a Coors commercial, that tumbles down the Bitterroots, Adirondacks, or the foothills of the Blue Ridge Mountains. That's a "B3" channel. Knowing these stream types allows you to begin to see what really defines these channels. It allows you to compare apples to apples and to understand what contributes to their differences. Most important, in answering the pressing need of those responsible for repairing streams, it provides specific guidance on how to restore these streams.

The system is somewhat intuitive but not something most newcomers would understand at first glance. The measurements are critical, but what exactly

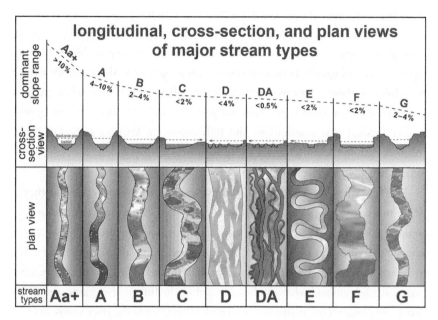

1.2 Graphic showing the typical plan, cross section, and longitudinal perspectives for various stream types as described by Dave Rosgen.
Courtesy of Wildland Hydrology.

should be measured is less clear. In streams, points of measurement are fluid in origin, moving their relative position when you turn a meander bend and head one hundred feet downstream. The mercurial channel widens without warning, then scours deep pools seemingly on a whim. This impermanence, not unlike the sand that forms many of these benchmarks, can create confusion. For anyone who has spent time in streams, it is clear that classes would be helpful.

In 1994, Rosgen published a seminal article in *Catena* titled "A Classification of Natural Rivers." This article laid out the system that he had been working on for approximately twenty years. Two years later, Rosgen self-published his book, *Applied River Morphology*, which fills out this system in color. The book, published specifically for his classes, describes the stream types; presents typical ratios, slope, and sinuosity ranges; and has diagrams that reveal key channel characteristics. It also contains a lot of pictures, riverine centerfolds that connect the classification to real world examples. But it doesn't explain how to use the classification system. You have to take the class for that.

Rosgen explains that he doesn't want people to use it incorrectly. This is a valid concern. Men and women without any training in hydrology, geology, or ecology who may now run a public works department are responsible for coming up with an ecologically beneficial restoration project. The short courses fit the schedules of those who can't stray from the office for too long. They are also a welcome escape from timesheets and performance evaluations. It's an opportunity to shed past assumptions and fully engage in learning something new. A chance to not be in the driver's seat. Rosgen is willing to lead the charge, but more than anything, he wants people to hear it from him directly.

People want to hear it directly from Rosgen. Quickly they learn that they'll learn more than just what angle to set a cross-vane. Rosgen provides an understanding of impacts that affect streams. He provides knowledge of how water shapes the land and how certain structures can guide the flow of water. It's training that provides a deeper understanding of what has happened to their particular stream and what can be expected in the future. It elevates everyone's work. As Rosgen says, it allows one to understand "what the river wants to be."

Half of our weeklong course takes place inside a comfortable classroom. After our extensive day of surveying, it feels right to settle into some book learning. This will be unlike any other lecture we've experienced. It's a time to sit down, get caffeinated, open *Applied River Morphology*, and listen to the creator of this system give us the good word.

Rosgen has more than a few words to offer over the course of the day. Despite the nearly eight-hour monologue that transpires over the day, surprisingly no one appears bored. Rosgen is a natural communicator and keeps our mixed group more than engaged. This is a challenge given that our group includes engineers, government managers, ecologists, and biologists all at different points in their careers. Rosgen keeps the group motivated with a series of slides that show various streams and rivers ranging from untouched to battered. This is why we are here—to fix these streams that have been damaged. We are in this together.

Rosgen seems to understand this, and his style is both inspired and down to earth. His cowboy bravado wins us over, yet he suffers no fools. Rosgen sprinkles in regular doses of folksy truisms that encapsulate a range of unwise decisions others have made regarding rivers. "Someone's got a terminal case of the dumbshits," he chuckles after showing how an attempt at pouring concrete over a bank to stabilize it led to the eventual collapse of the entire mess. In time-lapse slides, we see how upstream and downstream from the "fix" stream banks calved into the stream like glaciers breaking into the Arctic Ocean.

After showing a failed installation of an armored rip rap wall that was doomed the day it was installed, he chides, "That's like crapping in your chaps and sitting in the saddle." This chestnut of cowboy wisdom serves as an oft-repeated coda that follows Rosgen's examples of misguided endeavors. The public acknowledgment of these failures offers some comfort to the audience. Even if we miss a few of the details, we only need to avoid the inane mistakes of these unnamed simpletons. Who could be so stupid?

Photos of stream failures also keep everyone in line. From the assembled, there are only softball questions. Most of the attendees have seen failures like those shown in the slides. A certain number have probably had some involvement in something similar. There is an unspoken acknowledgment that we've been doing things the wrong way. Time to come clean. Listen to Rosgen.

Although Rosgen doesn't browbeat his audience, he does drive home a larger point throughout his lecture. Our collective attempts at fixing streams have been all patchwork and no process. Shoring up stream banks with all manner of stones, concrete blocks, or whatever else a Department of Public Works crew had on hand were dingbat actions. All of these half-assed attempts treated these issues as isolated problems. No one was considering the transport of sediment.

Over the next couple of hours, Rosgen mentions the term *dynamic equilibrium* at least two dozen times. All stream problems stem from a stream that is either aggrading or degrading. A degrading channel is one in which the energy of the stream exceeds its channel geometry, and the stream acts to reshape the channel to accommodate this energy. Streams can do this by cutting down into the underlying bed material, creating deep canyons and high banks in the process. Streams can also push laterally into their floodplains, creating a wider channel in the process. An aggrading channel is the flip side of this and can occur in the same river. The problems here may seem less severe but are only crises delayed. Stream energy drops as the channel widens and sediment falls out in apparent exhaustion. What follows are massive sandbars that will cause the stream to carve new channels into adjoining property within months. Sediment may drop out in oversized culverts, ensuring that future storms will flood the nearby road. What is needed is a dynamic equilibrium in which the amount of sediment carried in the stream is effectively passed through the channel. It is only when this state is achieved that the channel is effectively stable. Creating the proper channel geometry is exactly what Rosgen is going to teach us how to do.

Over the day we break it down. We get back into our field groups and look at the data we worked up at the bar. The reams of data points have been translated into numerous dimensionless ratios that we compare to the charts for the streams: sinuosity, slope, entrenchment ratio, bankfull width, pool to riffle ratio. These terms all describe different aspects of the stream reach we examined. My group compares our ratios to those of the other groups. By and large, the measurements align with the expected ranges. When one strays outside the range, Rosgen provides an explanation that the stream may be in a period of adjustment. The simplicity of the classification gives everyone a jolt of power. It all makes sense: C4.

Rosgen returns to his presentation to showcase some of the successful restoration projects he has designed. There is a picture of Rosgen holding up a brown trout in a recently restored stream. The river flows around him, guided in sine curves by immovable J-hooks and cross-vanes, stone structures that were placed to direct the water and create this specific pattern.

Rosgen brings up a project that had the right structures but missed a key measurement. He describes how this project failed after three years because the designers identified the bankfull elevation six inches lower than they should have. Comparison with a gauged reference stream would have prevented this. That difference equated to a twofold increase in stream power that overwhelmed the structures and created overwhelming bank stress; a million dollar project now needing a massive influx of dollars to repair. Finishing Rosgen's thought for him, several disciples start in on one of Rosgen's favorite bromides. As they chime their sing-song refrain, I see a few other students send playful grins my way: "If you don't know bankfull, you don't know shit."

THE BLOWBACK

Success draws attention, and critics soon began to sharpen their knives. Academics looked at this man without a PhD who was providing training on how to understand river processes in one-week increments. He was teaching more than two hundred students a year across the United States in classes costing $1,700 a pop that filled up shortly after they were posted. His classes drew people who were engaged, in some manner, in the messy process of restoring their rivers and streams. Rosgen's "natural channel restoration" gave these people a clear guide for restoration that sidestepped decades of research by academics. His approach had been blessed by Luna Leopold, one of the most respected figures in geomorphology, and it had the overarching goal of restoring a natural function to streams. Rosgen was helping people avoid the gross errors of the past. He was setting this developing field on a new ecologically friendly path. What was not to like? But to many academics it seemed as if Rosgen and his devotees were saying, "Keep doing your research; but we're moving earth here, and we'll let you know when were finished."

Rosgen had also made inroads with government staff who were looking for standards by which to evaluate restoration proposals. Government staff across the country had seen epic failures when permitted projects effectively eliminated habitat even if they survived the first major flood. His classification system provided these government staff members with a needed level of certainty. They could be assured that the end product would be both stable and would create habitat. It provided a sense of certainty that costly past mistakes wouldn't be duplicated. To ensure that only those with this mindset could design projects, government staff inserted requirements into RFPs that applicants must have Rosgen stream-restoration training.

Consultants who had designed many restoration projects found themselves disqualified if they didn't take a Rosgen course. University-educated geomorphologists who had spent hours modeling sediment transport of streams were not considered properly trained to do natural channel restoration, and to the ire of their professors they were regularly not hired for key positions managing this work. Even if academics who were experts on any subspecialty related to streams were to take these courses, which very few did, they quickly understood that they didn't fit into this new methodology. The myriad fields of study encompassing streams and rivers were all subsumed by the immediate question of channel morphology. What was the channel currently doing, and what did the channel want to be? The answers to those questions were determining when, where, and how streams would be reconfigured. Perhaps most important, they dictated who would be in charge.

The criticisms from academics ranged the gamut. A fundamental one was the question of the ability to reliably identify the key feature of bankfull elevation.

Relatively clear in most stable streams, this feature is much harder to identify in certain stream types with a coarser median bed material (think steep mountain boulder streams) and particularly in streams actively changing their form. These could be rivers that are forming anastomosed channels after being clogged with sediment or incised urban streams that have developed into small urban canyons gushing with stormwater flows. Every dimensionless ratio depended on accurate identification of this key elevation, they pointed out. If you couldn't accurately identify it in streams that were candidates for restoration, what use was this system really?

Another major criticism related to Rosgen's claim that natural channel restoration resulted in a stable channel. The payoff of knowing what the river wanted to be was that you could create a channel that didn't move and threaten whatever property or infrastructure it was currently threatening. Many academics pointed to streams deep in the Amazon that had been changing form for hundreds of years. They said that streams were by default systems that moved. They scoffed at the notion that an approach that called for a specific "correct" form would last even five years. They ridiculed in-stream structures created with large boulders that were deemed "natural" even if no such boulders existed in nearby streams. Perhaps most damning and revealing, they labeled Rosgen's approach a "cookbook approach," a knock against the perceived simplicity of the solutions.

Another common swipe came in less direct form but perhaps carried more professional angst than the others. Rosgen was making a good living with this training gig, and significantly more money was being spent on stream restoration since his natural channel restoration approach had been established. An influential article published in 2007 by Mattias Kondolf, Martin Doyle, Andrew Simon, and others laid out several key flaws of Rosgen's approach. Also included is a pointed statement that, up to 2007, the full expenditures on tuition and travel for Rosgen's classes were estimated to be between $28 million and $40 million. More than fourteen thousand people had taken his classes.[2] It's safe to assume that this exceeded the number of all people enrolled in all geomorphology programs in the entire United States over that time period.

The costs for these projects were raising eyebrows as well. Not only were individual projects regularly exceeding $3 million, but the number of projects was increasing exponentially. In the decade after publishing *Applied River Morphology*, the number of projects in the United States went from less than 400 to more than

2,400.[3] A 2005 meta-analysis by Emily Bernhardt and Margaret Palmer found that only 10 percent of project records indicated any type of monitoring, even though the annual average expenditure for stream restoration in the continental United States exceeded $1 billion.[4] It was clear that this field was rapidly advancing even though no one knew where this river of money was taking us.

WHAT WOULD ROSGEN DO?

Many years before my four weeks of Rosgen classes, I unknowingly began my circuitous path toward understanding Rosgen's system. I had taken a seasonal field position with the U.S. Forest Service immediately after finishing graduate school. I wasn't ready for a desk job that seemed predestined; I wanted some more time in the woods. In an unconventional onboarding process, I was told to report to a house the government had rented in the Catskills of New York and was paired with another recent graduate. The two of us were guinea pigs used to test the ability of future seasonal staff to classify streams using the Rosgen system. Although we wouldn't have the time or the budget for an official Rosgen class, the Forest Service flew in two seasoned hydrologists to teach us. They had both taken the classes and understood it well. We were both completely new to the world of stream classification but had a strong desire to learn something new and get paid to walk around in streams.

Our unofficial Rosgen training got started right away. We were given waders, and our teachers drove us out to a nearby stream. Immersion in this work is both literal and figurative, and we enjoyed both. To begin, we waded one of the thousands of clear Catskill streams stabilized by the quartz boulders and gnarled roots of hemlocks, which were captured precisely by the artist Thomas Cole. Ours was a deceptively rugged stream, not full of waterfalls but dozens of minor boulder-filled cascades that tempt you to cross them and twist your ankle in the process. The stream seemed to be in an active battle with these boulders carved from the Devonian bedrock that comprised these mountains—each vying for preeminence.

Back at the rented house, our trainers used simple diagrams with varying perspectives to help us see the river in a different light. Cross sections, longitudinal sections, plan view; these diagrams broke down a river in components that could be measured. Pools, riffles, glides, runs . . . these components were all familiar to anyone who had ever walked in a stream. It made intuitive sense that they would form in relatively predictable patterns. Moving downstream, riffles transitioned into runs that transitioned into pools. Glides were the upward-sloping features as you moved out of a pool. Identifying and naming these discrete components gave us a sense of control over these otherwise unfamiliar rivers. Our task was to identify and survey the elevation of the exact beginning, ending, and all-important intermediate points of these features. We were well trained and felt empowered. We would gather all this information on our data sheets, take them back to the house, and work up the data. After calculating numerous ratios, our hard-won measurements would be transformed into a satisfying and data-derived alpha-numeric classification.

The process was the riverine equivalent to a doctor performing a detailed physical. The 350-foot section of stream would be carefully assessed and categorized. There was a palpable feeling of discovery that buoyed our spirits. Our particular channel had not yet been classified, so we were defining something for others in a way that could help people understand what was happening in this stream. If we weren't performing life-saving surgery, we were at least diagnosing the patient and setting a baseline for a future of healthy living. Much like getting a physical with your doctor, it was time-consuming, sometimes uncomfortable, and inarguably necessary. We laid the channel out in our plan-view sketches, detailing the bends and major features: a point bar here, high banks along the channel just downstream, an eddy forming behind the roots of a tree that was anchoring the bank, at least for the time being. With the enthusiasm of an art student in Florence, my partner meticulously sketched in a nice stand of ironwoods that anchored the right floodplain halfway down.

Having established the lay of the land, we picked out three riffle sections where we would survey cross sections spanning the channel and up onto the floodplain. This would provide key measurements of channel width and depth. Perpendicular to these cross sections, we surveyed the entire 350-foot length of our stream reach in what was called a longitudinal profile. Starting upstream and heading down, we followed the lowest point of the stream channel, trying to catch the subtle break points where gravel riffles would change slope and turn into the beginning of a pool. Sensing these undefined points invisible below the water surface required either an uncanny sense of elevation or a certain faith in our abilities. Does the top of the pool start here? Perhaps two feet upstream? Questions made more difficult by the distance between me and my partner.

The only way to put the questions to rest was by taking repeated survey measurements—never too many measurements.

Although all points of elevation had significance (not finding the lowest point in the pool would mean that the measured depth was less than the true depth), one measurement loomed over all the others. This was a measurement we couldn't mess up—the bankfull elevation. Like identifying a pulse, it set the baseline for all other measurements. This elevation was a feature that had specific hydraulic implications, and its manifestation was a physical indicator that could be identified with some training. Not unlike finding a pulse, it took a bit of trial and error. Just like moving from a wrist to the jugular, in some cases there were better places to measure it. Similar to whether you timed out fifteen seconds on your wrist or twenty seconds on your neck for your pulse, there was only one true bankfull elevation, and every diagnostic that followed depended on an accurate determination.

The best way to envision bankfull is to imagine a canoe trip down a lazily meandering river. At some point, a sandbar presents itself on the inside bend. Perhaps you pulled your canoe up here so you could take a break and take a quick dip in the pool on the other side. If you were to walk out of the river proper and up the sandbar, the point at which the sandbar levels out would be bankfull. This feature is critical to the classification systems because all of the measurements are based on it. The bankfull width of the stream is the width from bankfull on one side to bankfull on the other. The bankfull depth defines the depth as the difference between the elevation of bankfull and the deepest point along a transect across the river. In this way, changes in the position of bankfull change all measurements and the associated ratios that define the channels.

Back at our cross section, we take many measurements but an uncomfortable sense of uncertainty creeps in from the dense forest. This key measurement eludes us. There are no sandbars calling for us to picnic. In fact, there is no sand at all except small pockets lodged in between boulders. We take a survey reading anyway. One small flat area ahead reveals itself but measures only two inches on each side. It seems too diminutive a feature to stake our claim of bankfull, but we take a reading anyway. Another is larger but has formed behind some tree roots in a scoured area underneath a streamside hemlock. The protection of the hemlock roots must invalidate this indicator and makes it nearly impossible to maneuver the survey rod into a vertical position. I ask Ryan for the elevation readings of our potential spots. They are within a two-foot range—which is sort of like saying your pulse is between twenty and two hundred. A two-foot difference in stream elevation translates into a twentyfold increase in flow volumes. We rot in the insidiousness of bad data.

This can be sorted out at the office we tell ourselves. The truth is that the uncertainty haunts us. Did we pick the wrong area to set up the cross section? Has a recent storm pushed bankfull indicators downstream? Is the dense vegetation covering up the right indicator? What about those boulders that sat at the base of a stream bank and appeared to have been pushed down by someone or something in the recent past? WWRD? We plod on, coaxing the mind-numbing beeps out of the survey rod and hoping that all will be revealed at some future time.

SHINING A LIGHT ON AN INTERNECINE FEUD

"Now they are not only attacking Dave's methods, they are attacking him personally." I received this terse and wary email from a respected colleague very familiar with Rosgen's system. Attached to the email was a copy of a dissertation of a relatively unknown PhD student. I wasn't quite sure what to make of it, but the juicy nature of the message piqued my interest. Someone was going after our guy, the man who had shown me how to understand rivers. Although I wasn't in the habit of reading dissertations, this one grabbed my attention and didn't let go.

This was how I first heard of Rebecca Lave. Lave was an anomaly in this battle among river jocks. She hailed from sunny California, was just beginning her academic career, and was a woman. Despite this, she quickly found herself at the center of this controversy. Her spotlight on the inner workings of the field would expose many red-hot issues and would quote many prominent people in the field. People wouldn't see things quite the same again.

Her interest started when her work drew her into a stream-restoration project. Lave was an environmental consultant in northern California working on permitting for a stream-restoration project near El Cerrito. In the process of talking to the restoration designer who was tasked with restoring this suburban stream, she was struck that the designer chose to jettison a recommended meander bend to avoid disturbing a small restoration project built by the local "Friends of Five Creeks" group. This "friends group" had not chained themselves to the trees or even raised any red flags, but the prospect of ticking off an influential group was scary enough that the designer dropped his initial plans. Puzzled that a scruffy grassroots group could have such outsized influence, Lave soon signed up for the geography program at the University of California, Berkeley.

With this experience in stream restoration, she quickly ran into Mathias Kondolf, professor of landscape architecture and environmental planning at Berkeley. Kondolf was a highly regarded fluvial geomorphologist who introduced students to the numerous interconnected aspects of river-restoration efforts. He had studied under the legendary Reds Wolman at the Johns Hopkins University and subbed in for him as needed. He was a vocal Rosgen critic and spoke point-edly about the numerous failed stream-restoration projects he had witnessed. He was known for sending his graduate students out to monitor streams after restoration to the annoyance of consultants who designed the projects. Some consultants referred to these graduate students as "Matt's student hit squads" because the data they collected ultimately resurfaced in blistering critiques of these projects at subsequent conferences. The common thread in all of these projects was that they used the "natural channel design approach." These were restoration projects done with Rosgen's methods, empowered by Rosgen's train-ings, and likely awarded due to Rosgen certifications.

Kondolf also called out the consultants directly. During one plenary session at a northwest stream-restoration conference, he stopped his presentation to place one consultant's business card on the overhead. Channeling the ire of all aca-demia, he pointed out that the professional claim of geomorphologist, rather than being achieved by any rigorous academic training, could now be met by being "Rosgen level-4 certified."

Lave knew that the rule of thumb in social studies of science is that you go where the controversy is. It was clear that the subject of her PhD research pos-sessed that element. Yet rumors and anecdotes are not material for a dissertation. Kondolf wanted to help this new graduate student and in 2003 found some left-over grant funding to bring Lave along to a National Research Council meet-ing in Minneapolis. This conference was organized by several people who had heard snippets about the evolving controversy and wanted to get those involved together to talk. The goal was to develop a report that would summarize agreed upon principles that they hoped would come out of the conference. As Lave explained, there was a desire from the organizers for everyone to "make nice" and figure things out. Naturally, the organizers invited Dave Rosgen.

The conference kicked off with a plenary session highlighting stream-restoration efforts. Kondolf arrived at the session late but didn't hold back his fire. His critiques of natural channel design were sharp and aggressive. Many in the room felt uncomfortable, and this combative atmosphere continued throughout the conference. Despite the attacks on him, Lave noticed that Rosgen, dressed as if he had ridden his trusty palomino to the conference, remained unphased. Despite the battle lines that were being drawn, he was not finished trying to win them over. Further presentations and flare-ups continued throughout the conference, and the organizers never got their report on unified approaches.

The conference provided Lave with a close-up perspective of the people behind the controversy. This controversy would soon become known as "the Rosgen wars," and it was fundamentally a battle for primacy in this newly developing field. Overwhelmingly male and overwhelmingly white, the combatants knew the implications of these discussions. The battle took on a viciousness that has guided alpha-male battles for millennia. The fight was not only for technical and scientific primacy but also a battle for influence in determining how and where funding should be spent to restore rivers. And perhaps most personal, it was a battle to determine who would be given the honor of overseeing this work and the future renown that would likely follow.

Armed with ample material and certain that she was researching something significant, Lave left the conference excited. Living on a graduate school budget, she decided to share a cab to the airport. In a fortuitous turn, the man in the taxi line ahead of her happened to be Rosgen himself, and he gladly agreed to share a cab on the ride to the airport. After learning that she was a graduate student, he asked what she was studying. "You!" Lave admitted with a forthrightness that impressed Rosgen. This began a long-term friendship, and Rosgen provided extensive access to his background, his career, and his training. In future classes, he would proudly mention that he was the subject of a dissertation being produced by a young woman from Berkeley.

Lave began the extensive social science research of reaching out to more than one hundred people involved in the stream-restoration field. She found that people were eager to talk. Consultants who had been working for decades in the field felt shut out of work for not having participated in the Rosgen training. Some consultants gave up bidding on any government-funded projects knowing that their rigorous academic training would not qualify them under the terms of carefully worded RFPs. Incredulous academics talked of harried attempts to form cross-subject teams to investigate the growing number of stream-restoration projects. Their judgment, carefully worded but usually critical, trickled into the scientific journals.

Lave also spoke to many Rosgen supporters. Many government staff members appreciated that Rosgen provided an easy standard by which to evaluate a constant stream of consultants that claimed their authority. They bemoaned a "lack of credentialing or even consistent academic coursework sequence that one could use to screen out people who have skills and those that clearly don't."[5] Rosgen provided some guardrails in a field that was becoming the wild west of environmental consulting. Even some academics appreciated what Rosgen was doing. They described an academic community that had been telling consultants what they couldn't do for decades. Gary Parker, a prominent researcher from the University of Illinois, stated: "Rosgen has had the effect of moving the entire field of river geomorphology more in the direction of thinking about how to solve practical problems."[6]

Harder to document were the feelings of attendees at his short courses. Lave attended two courses in 2006 as a stated observer. Nearly all adored Rosgen, but no one wanted to speak on the record. They felt lucky to be under his tutelage and were empowered by the clearly applicable knowledge he was providing. Rosgen did more than share a useful classification and needed framework for restoration: he provided a sort of American optimism and purpose that outshone his critics. Compared to the normal gloom and doom that pervades much of the environmental field, Rosgen's can-do attitude fed their souls. Students breathed in this optimism, dutifully learned his classification system, and resolved to fix the stream problems that were waiting for them. Students became lifelong friends, and networks of the Rosgen-certified developed organically.

After completing the interviews, Lave produced a particularly insightful account of the multilayered influences of this new approach to stream restoration led by one charismatic individual. Many observations flowed out of the report. The academic community wasn't providing the type of training and guidance that private and government sectors needed. The academic community was circling the wagons, dropping prior paths of study, and forging research groups to squelch these out-of-control practitioners. Government staff were quietly inserting key phrases in procurement documents to ensure an outcome with which they felt comfortable. Ambitious consultants were running with the template that Rosgen provided and growing their businesses and the overall field in the process. When Lave shared it with the people she interviewed, she heard from nearly all that it accurately described what they saw happening in the field. For the social scientist, this is the gold standard that proves you have developed a useful analysis.

A gold standard sounded good, but she was less prepared for the unprecedented interest her paper received. Lave said, "For most dissertations you're lucky if your committee and your mom reads it." She soon found that her paper had been posted on general listservs and was being emailed around in the restoration network. She said this unexpected and often anonymous circulation of her findings to an unknown audience that was primed for a food fight was not a comfortable moment. Through emails, classroom lectures, and sidebar conversations at conferences, people had been passive aggressively sniping at ideological enemies for years. Now a researcher had described the entire dynamic in a way similar to what a family therapist might do in shining a bright light on a never-discussed family dysfunction.

Although Lave's account adhered to all academic protocols, the subject matter dripped with juicy details for those in the field. This was personal. Consultants well known by their peers shared their fears about irrelevance and financial ruin. Respected academics shared their thinly veiled plans for a counterattack. Government staff shared their frustration with identifying the right approaches and ensuring a proper use of public funds. But most of all, the paper was fair to all. Rosgen was given credit for his training and accomplishments, and the key criticisms of his most fervent critics were presented in an unbiased fashion.

The subject of the controversy seemed to take it in stride. Lave states that "Dave took issue with a couple of points but overall was not upset." She received one pointed critique, but mostly people agreed with her analysis. They saw themselves in the controversy, but this self-realization didn't translate to any widespread comity. "By the mid-2000s, most academics assumed they lost this battle and gave up," Lave states. Academics would continue their research, and Rosgen-trained practitioners and Rosgen-trained government staff would continue their work of fixing streams the best way they saw fit. The two worlds existed in parallel but distinct realities. "What's interesting about this issue is how hard it is to resolve," Lave concludes.

Our last task is to assess the streambed material. Wolman's pebble counts are an assessment that seems ridiculous until you understand it. I pace across the stream at a set distance (a single stride) and then measure the pebble, stone, or boulder that I first touch when reaching down a finger, eyes closed, to the streambed. There are several sides of a rock, but I choose the one that would be the limiting dimension in a sieve. I continue this sort of Frankenstein-inspired walk, pacing and stopping until I hit the bank and turn around. Pace, bend down, measure, repeat. We can stop when we have one hundred entries. Sometimes you measure rocks because you are paid to measure rocks.

This being our first jobs out of school, we both want to get it right. But self-doubt begins to seep in. Did I really pick that stone, or was I drawn to it due to the fact that it was slightly higher perched relative to the one next to it? Am I picking all small pebbles because subconsciously I'm too lazy to pick up the boulders? Am I avoiding the really round ones because it would be confusing to know which side to measure? Do these questions cloud Rosgen's mind?

As we continue the transects, me pacing the stream and my partner writing down diameters, two women from a nearby camp walk by and assess the availability of the swimming hole. "You guys counting rocks?" "Counting and measuring them," we correct. They make no attempt to hide their laughs and move along. Apparently swimming and Wolman's pebble counts aren't compatible activities.

PULLING NO PUNCHES IN ADVANCEMENT OF THE SCIENCE

Rosgen still offers his short courses today. Consultants still take the classes. Many talk of using some of Rosgen's methods but not adhering to his approach in lockstep. They know Rosgen, but they're not "Rosgenites." They talk about building on the Rosgen approach, making sure to establish their intellectual independence. The clarity and purpose that Rosgen inspired in so many fades into tepid endorsements and approaches full of conditionalities. Rather than following a script that unifies action, practitioners insist on their own specific approach that, not surprisingly, requires the expertise of that practitioner.

As for the academics, there have been attempts to create an alternative. Kondolf is at the forefront of these efforts. With a black shirt and glasses, slicked back gray hair, and a gravelly voice evoking a "C4" channel riffle, Kondolf exudes academic heft. I sense conviction, free of bullshit and politically couched terms, but his conviction is rooted in the power of science rather than personality. The giants of the field I have researched online he learned from directly. He subbed in for his primary advisor Reds Wolman. His current position at the University of California, Berkeley is the same one once held by Luna Leopold.

Bucking all of my conflict-adverse tendencies, I uncomfortably ask Kondolf about his testy reputation in the ongoing dialogue about stream restoration. Although opinionated and principled, he also strikes me as a decent person. "It's true that I've been very critical at times," he offers with measured reservation. It's left unsaid, but I intuit that he might have some misgivings that feelings have been bruised over the years. He continues, "It's true that I regret the investment in natural channel restoration. It's a rathole we're pouring money down that could be used for a lot of other things." My armchair psychology was apparently off base.

Kondolf has not been idly stewing over this flow of money toward ill-advised projects. He tells me about a short course he has been teaching with several colleagues for more than twenty-six years. The RiverLab offshoot of the UC Berkeley Landscape Architecture Program offers weeklong classes in Lake Tahoe and in Lyon, France. The class in Lake Tahoe provides consultants, agency staff, and some academics with a combination of lectures and field experience in a beautiful setting. Students meet in rustic classrooms, sit on logs around campfires, and wade in tranquil streams that feed the azure Lake Tahoe. When asked what the attendees are qualified to do after this course, he offers, "We don't provide the keys to the car. The goal is to understand the underlying process and the underlying science." The course fees of $2,400 include all meals and associated materials.

I ask him if he's ever taken one of Rosgen's courses. He relates that at one point in the past he was informed by an intermediary, one of Rosgen's staff, that he had been "signed up" for a course—tuition free. He was unable to get away from his classes and take Rosgen up on the offer. I'm struck by the missed opportunity and the universally poor communication between members of the male sex, regardless of rank or station. What might have happened if everyone sat around the same campfire?

I ask if we are getting anything right in the stream-restoration field. Kondolf points to the fact that he is seeing more process-based stream-restoration approaches, especially on the West Coast. He applauds dam removal efforts and efforts aimed at removing miles of remnant concrete channels, most notably in Milwaukee but also in nearby Contra Costa County. He mentions that some newer dams in Switzerland and Japan are threading the balance between providing cheap and carbon-free electricity while allowing sediment to pass through the dam. It's clear that he's feeding his students these positive examples as well.

Still fumbling with my conflict aversion, I gently inquire if, as Lave had suggested earlier, the academic community has essentially given up on the fight with Rosgen. He pauses, breathes in deeply and then exhales a breath mixed with resignation and frustration. "There is some truth to that statement," he acknowledges. "You can't keep fighting this thing all the time." Agency staff still cling to the natural channel approach, and Rosgen is a charismatic person, he adds. Yet he doesn't seem content to fully admit defeat. Kondolf mentions that he and several colleagues are developing criteria for process-based designs. These criteria, with their deep and full academic consideration will certainly be surfacing soon, not so gently nudging those spending the millions to think once more before the contracts are signed.

———◆———

Almost fifteen years later, my memories of Rosgen resist easy categorization. There is the residual sense of uncertainty that I felt conducting stream surveys. Maybe our data would have been better if we had had more access to Rosgen. Yet those errant data points lose significance over time. What I'm left with is more lasting: the sense of purpose and clarity remain. In the Anthropocene, the epoch in which the earth and its numerous processes are increasingly shaped by human activity to the detriment of so many other species, the key questions center around what humans should do. Few have the confidence and panache to provide the answers. More than the ratios, I remember the fleeting moments at the end of the course: Rosgen shoehorned by a dozen participants with specific questions about their rivers, still wanting to talk after five ten-hour days of river talk. Rosgen making suggestions to collect more data. Saying he'd be happy to look it over. Armed with field books, the *Applied River Morphology* textbook, and a new way of looking at rivers, we were empowered. We could do something about fixing the rivers we loved.

CHAPTER 2

THE BOG ARCHITECT

Reimagining Streams and Stormwater on the
Coastal Plain of the Chesapeake

I drive into a residential neighborhood, get out of the car, and slip into the forested park across the street from a row of middle-class homes. Immediately I descend a steep slope whose bottom lies some fifty feet below. I shimmy down, taking care to avoid loose logs hiding in the leaf litter. An errant step might lead to a tumble down into the creek far below.

Two-thirds of the way down, a precarious bridge extends unsupported over the stream. Two feet wide and made of concrete, it has no rails to prevent anyone from falling into the newly formed eddy below. Although the stream is only a couple of inches deep, it has carved out multiple channels across the bottom of the valley. Following soothing gurgling sounds, I walk out onto the unlikely bridge to investigate the firm footing it provides. A recognizable odor settles in, and it becomes clear that the gurgling sound is not the stream ten feet below but a river of sewage from the homes above. I'm standing on a concrete encased sewer main. I slowly realized that this unsupported twenty-five-foot section used to be buried. I decide to get off the pipe before it collapses and unexpectedly spills its contents on me.

The stream is barely deep enough to overtop my running shoes, so it cannot sustain fish. There are no deep pools, and the water is cloudy with sediment. A sudsy foam collects in the eddies. There are no insects skimming the surface and no aquatic plants growing in the channel. Although docile today, there is ample evidence of the stream's restlessness. Unsatisfied with its already ample channel width, the stream has pushed into the valley wall. Uprooted trees span the channel at awkward angles, providing no certainty as to their stability. Buried in the hillside ahead, a massive, four-foot-high concrete outfall discharges a trickle of water over a greenish slime that covers the base of the headwall. Other signs of suburban detritus include rusting bike tires, bloated basketballs, and the odd potato chip bag.

Upstream, vertical banks reveal a clear layering of soils. At foot level, a layer of rust-colored sedimentary stone has halted the stream's downward incision. At eye level, a layer of smooth, consolidated clay creates a gray horizontal band that parallels the stream. Before its recent exposure, this clay had been resting underground since the Cretaceous period—some sixty-five-million years ago. Above, cliffs of unstable, red-streaked sand extend twenty-five feet up toward the sky. These ancient deposits are now exposed to the forces of water, both liquid and frozen. If I could time travel back one hundred years and retain my exact position, I would be buried up to my neck if not completely covered, a geologic erratic warning of a future era of mass erosion. What was laid down over millions of years has been unceremoniously flushed downstream in a couple of decades.

The houses at the top are no longer in sight. Looking up from these depths, I see ominous rain clouds slip through the fingers of barren tree branches. Time to get to higher ground. The scramble up is filled with several slips and numerous curses. At the top, I sit on the curb to catch my breath. As I pick mud off my running shoes, I see that I've brought with me all the layers of the local soil profile.

This precarious situation in the valleys betrays the quiet, leafy calm that this neighborhood in southeast Washington, D.C. exudes. The stream is a tributary to the Anacostia River, which has suffered decades of neglect despite being within a mile of the nation's capital. The lack of attention extends to this tributary; it hasn't even been given a proper name. Deprived of the rich linguistic options of run, kill, sough, arroyo, or fork, it is known by the name of the park: Alger Park. In what would be a fitting revenge, it may soon devour its namesake. But this scene could also be filmed in any tributary to the Severn River or the Magothy River across the border in Maryland. To those who are paying attention, this movie is quite familiar. But what is the backstory? How could a stream barely larger that a trickle create this gorge?

Flash back to the time of expansive suburban development, beginning somewhere in the 1950s and extending into the 1970s. Rather than focusing on the wood-paneled station wagons and the father in the gray suit, the camera pans down and follows the curb. Endless miles of driveways, culs-de-sac, and roads stretch to the horizon, all connected with a curb and gutter system that swiftly delivers rainwater to the nearest catch basin. This prosaic scene masks the real crime at the end of the pipe.

Throughout the Coastal Plain, an area extending from Delaware to Georgia, the tangible impacts of development are being exposed. Several decades after the push to build our urban and suburban neighborhoods, the ground beneath us is literally giving way. The transportation network that connects us is doing more than just providing safe passage home. This network of hard, impervious surfaces delivers vast quantities of rain, or stormwater runoff, down through catch basins and into a network of stormwater pipes that end at an unadorned pile of stone along the banks of the closest stream. This arrangement was relatively inconspicuous and worked for some time. Moving forward in time-lapse fashion and adding a pattern of increasingly intense rain events, our actions are now catching up with us. Simply put, this discharge of stormwater has left us with a problem that most of the public is unaware of and few in government want to claim, assuming they can even navigate the terrain to witness the damage.

KEITH UNDERWOOD'S COMPOUND

There was no direct communication prior to our meeting. Despite steering every aspect of a busy twelve-person design/build company, Keith Underwood is not readily available through most modern communication channels. His cell phone message says not to leave a message because he will not check it. Through my previous job I had known Underwood for about fifteen years, so this wasn't an issue because there was no need for introductions. I knew I needed to talk with Underwood in person, so after exchanging some emails with his assistant, Heather, I scheduled a trip to Maryland.

We were set to meet at Underwood's new property that, over the course of our tour, would go by several names: office, shop, yard, and "farming operation." His property is located in a rural area not far outside of Annapolis, Maryland, where small, remnant hobby farms are mixed with larger estate houses for the 1 percent of Anne Arundel County. I had plugged the address into Google maps and was directed to the left, but when I saw an impromptu tree nursery and distinct sedimentary boulders with recognizable red and orange streaks on my right, I was certain I was in the right spot. Reminiscent of rusted iron on an abandoned piece of farm equipment, these stones are Underwood's calling card. They're known alternatively as bog iron, limonite, or ironstone. The stones are placed throughout the 8.5 acre property. As I drive up the sand driveway, a few stones have been placed next to a newly formed pond. Others have been arranged to form stone retaining walls bordering the long driveway.

A stately brick house sits resolutely at the pinnacle of the property. It's accessed by two opposing curved driveways that form semicircles from above. As I drive up toward the house, the importance of this layout is not clear to me, but the slope and the curve create a grand entrance out of a Tolstoy novel. Large tulip poplars, maples, and oaks bracket occasional views of the house. It is a lovely piece of property that is currently being reshaped and reformed to fit a specific vision.

Underwood meets me in the driveway with a pirate smile, and we wander toward the crest of the hill to take in the view. A couple of his staff wander over to join us (figure 2.1). It's the crow's nest of his landship, and we survey the expanse before us for threats. He lights up and drags from a cigarette that's a ubiquitous fixture in his hands when any person approaches with questions. He's intent on showing me what he's been up to since we last met five years ago.

2.1 Keith Underwood, Chris Becraftand, and Keith Pivonski at the office/yard compound with a bog iron stash.

It's a beautiful sweep down to the small stream that follows and then passes under the road I drove in on. There is currently no evidence of a headwater stream on Underwood's land; nevertheless, he elaborates on his prime spot in the headwaters. In subterranean fashion, rainwater will be infiltrated, absorbed, cleansed, and allowed to seep into the newly created pond. The pond will supply the tree nursery below. Underwood speaks with such assurance that I can visualize these underground flows. He points out that our current position is at the top of a wind funnel and describes plans to put up a windmill to catch energy—wind energy to supply his crew's compressed air pneumatic tools. Underwood mentions that he has invested heavily in the property and that the upgrades to the house have been more expensive that he initially had thought. Nonetheless, he is confident that every aspect of his business is now under his control and that the various rental payments have ended, so the investment will pay off.

The reverie is broken as Underwood overhears a cell phone conversation one of his staff is having with some government type. Underwood butts in, "We've GOT a permit, it says it's a permit! Just go!" The staff member is doing his best to listen to two people at once. Then a minute later, "They know they can't put that shit in a fucking permit! Just go—do it by barge!" Another employee uses the break to ask about an email he sent. Softening to a pianissimo, Underwood responds, "Yes, a home-equity loan—that's what I said."

We enter the 150-year-old brick house that exudes a serene and bucolic past. The house is beginning a different era with the new ownership. Two of his project managers, KP and Chris, and his office manager, Heather, join us at a conference table. They are a youngish group who seem excited about their work. It's clear that Underwood's renegade personality and contagious conviction draw people who are passionate about ecological restoration. More than one employee left a stable government job with benefits for a less secure position with Underwood. More than one has left the job in tears. Those who stay understand that the work is paramount, and they will take some bruises if that is the cost of being involved with this work.

Most of Underwood's staff have studied ecology and understand the complex interactions that characterize healthy functioning ecosystems. Their desire to repair them is evident to me. However, these aspirations are more felt than stated, as if stating them could insert a level of human ego into an already unbalanced ecological equation. I know this because I have held similar positions and share their background and outlook. But understanding the problems and knowing how to fix them are two different skill sets. Like many other fields, diagnoses are plentiful and cheap, solutions are rare and usually expensive.

Any good ecological education is tempered with the precautionary principle, a subtext in all ecological teachings—first do no harm. It also emphasizes caution prior to jumping into solutions when scientific knowledge may be incomplete. Faced with a damaged river or stream, it's not always clear how to repair the damage that stems from multiple sources. Underwood doesn't seem to dwell on this principle. He knows what to do and proclaims his vision for a restored ecology with the conviction of one who has seen the light.

Given the state of our streams and rivers, this certainty is comforting. Most people who've been in the field for a couple of years have seen half measures. Those drawn to Underwood want projects that go beyond just establishing a specified channel dimension or required vegetated coverage. They want their work to be regenerative—a term used frequently in Underwood's universe— directly addressing and reversing a suite of anthropogenic impacts by starting a cascade of positive interactions that builds function out of a broken ecosystem. This involves much more than stabilizing a situation or restoring with proper vegetation. It means creating the ecological infrastructure that will build off the modified hydrological inputs and, over time, create a localized ecological system that thrives at the site, bringing with it the range of biota, ecological function, and stability. Brimming with aspiration, the term implies that prior restoration efforts might have acted only as Band-Aids.

The staff have brought out several bound monitoring reports that document specific projects. There has been a lot of research recently on the stream-restoration approach that Underwood has developed, some positive and some skeptical. They are a mix of journal articles, white papers, and self-published proceedings that circulate within a very small community. After years of challenges from regulatory staff, competing consultants, and some academics, Underwood relies on the positive reports when they go for permit approval. In a halting cadence, Underwood carefully explains that they need to get more research out: scientific credibility is critical. It is also uncontrollable and can originate from any quarter. "EPA had some interns doing some monitoring critical of one of our projects," Chris mentions. "They were completely exposed down at the Fish Shack. Run out of town," he says, referencing the regular meeting location in neighboring Annapolis for government types involved in the Chesapeake Bay restoration.

Underwood picks up one of the reports that maps all of the bogs in Maryland. "They're our natural analogue for our restoration approach. On a scale from 1 to 10, restore it to an 11." He mentions that the one thousand Atlantic white cedar trees planted at Howard's Branch doubled the population west of the Chesapeake Bay. This maximalist approach comes with more than a bit of eco-swagger. But the goal is restoring ecological function, so it's more inspired than boastful.

———◆———

While I scan through one of the reports, he tells his staff to fix the directions on Google maps that pointed me to the wrong side of the street. I can see how his attention to detail could be overwhelming. It seems as if Underwood's internal task list is constantly being updated and delegated. I ask if we can see the rest of the property.

We walk to the back of the property and look at the piles of bog iron boulders laid out in a scree pile along the dirt parking lot. This arrangement ensures that they can be easily picked through and loaded onto trucks. It's easy to forget that this is a working yard. Stones, soil, and wood chips will need to be accessed for different projects. Everything is in flux here, and continuous adaptation is the modus operandi. Underwood mentions that he has recently brought these boulders to this new property from several stashes throughout the county. They are heavy baggage but essential for his work.

He points out a fossil of a prehistoric horsetail plant in one of the boulders lying near the road. He says they try to save these fossils when possible. I'm don't know if they are rare or not, but their fossil history seems to reflect a sort of provenance that Underwood appreciates. Underwood had first found this particular type of stone at a development site and got them for free. They were not valued at the time due to their sedimentary and somewhat crumbly nature. Now Underwood is fighting for them because he knows that competing companies want them. He is paying more and is attempting to build a supply that will keep him in business for a while. His motherlode is growing, but they are still layered with conflict. "On the last project they wanted me to put these stones to a hardness test. Why? They're not supporting anything other than plants!"

Near the house these same boulders are set into the hillside and frame the entrance to the newly retrofitted walkout basement. Carefully placed bright green cinnamon ferns pop out of the gaps between the boulders. There is clearly a level of organization amid all of the earth moving. I ask Underwood how many years he's owned the place. "Three months," he responds.

ATLANTIC WHITE CEDAR PRIMER

Underwood is a creature of bogs. Not just any bog, but an Atlantic white cedar bog. Understanding Underwood first requires understanding these bog systems. Named after their dominant tree, these bogs were in abundance prior to the development and wetland drainage along the Atlantic and Gulf coasts. Occupying a narrow band that reaches no farther than 130 miles inland and extends from the panhandle of Florida to southern Maine, these trees are adapted to disturbance.[1] Assuming a local seed source, Atlantic white cedar can establish themselves following a storm surge of saltwater that kills a hardwood swamp forest. They are ideal buffers to coastal flooding. In addition to their wide temperature tolerance, they also tolerate wide ranges in inundation. Despite these adaptable traits, this wetland forest type has been reduced to only 2 percent of its historic acreage and is considered globally threatened.[2] They are currently the dominant species in only 170 square miles—less than a third the size of Virginia Beach, Virginia. Golf courses in the United States currently occupy 3,500 square miles. Compared with other wetland forest types, they also get the short end of the stick. The oak-gum-cypress forests that dominate the saturated soils of the southeastern United States cover a respectable 44,000 square miles.[3] In ecological circles, Atlantic white cedar bogs are a beloved supporting actor that receives adoration from those fortunate to be in the know.

As tree go, they possess fanlike, blue-green foliage and small blue cones. In the spring, they produce small red-yellow flowers. A champion Atlantic white cedar tree in Alabama reached eighty-seven feet tall and was estimated to be 258 years old. However, being susceptible to windthrow or fire, they rarely reach these ages.[4] Because of these regular disturbances, they are classified as a small-medium columnar evergreen tree, a somewhat modest designation for such a beautiful tree.

In this case, looking only at the tree is truly missing the forest. The benefits of an Atlantic white cedar forest originate from the dense habitat they create more than any particular characteristics of the tree. The dense thickets are perfect cover for otter, rabbits, deer, and bears. Deer are sustained throughout harsh northeastern winters by the tree, and many species of birds are suited for a cedar forest, in particular the prairie and worm-eating warblers.[5] In fact, one study in the Great Dismal Swamp in North Carolina found that Atlantic white cedar swamps sustained twice as many birds per unit area as the nearby maple-gum wetland forests. Seven bird species existed only in the Atlantic white cedar swamps.

Their benefits extend beyond wildlife. Coastal resilience planners are looking at Atlantic white cedar forests as a potential tool to address increasingly strong hurricanes brought on by climate change. Their ability to absorb the blows from natural disasters could potentially save our coastal communities, if there were any places left to put them.

———◆———

Atlantic white cedar forests require soils low in nutrients that are regularly inundated with water. Other species that need these conditions include poison sumac, red chokeberry, sweet pepperbush, bitter gallberry, fetterbush, and poison ivy; plants whose names reflect their ability to adapt to difficult conditions and their disinterest in being "team players." Possessing a less alarming name is the delicate, round-leaved sundew, a tiny plant whose fragility belies its carnivorous behavior. Armed with sticky glandular tentacles, it supplements its poor mineral nutrition from sandy soil with insects. Within each bog, each plant has a specific niche where it can thrive; this translates to somewhat predictable bands of vegetation. In the center of the bog, sphagnum moss creates a floating mat that can be thick enough to support a person. Thickets of high-bush blueberry find purchase on occasional hummocks that provide their roots with a growing medium, albeit one supported by water rather than bedrock. Close to the edge of the water but still completely saturated, magnolias and Atlantic white cedars form a denser ring of taller vegetation. Shrubs such as leatherleaf and clethera share this space with these trees. Moving outward, the soil surface becomes drier and the soils are sandy. At this time of the year, bracken ferns rust in the understory, still retaining their form. As the topography rises to the surrounding ridges, holly trees, pitch pines, and white oaks find their places. It's a visible progression of species, an ecological textbook manifested in living form.

HOWARD'S BRANCH: THE FIRST PROJECT

We beat the others to the site of Underwood's first major project and start down the steep slope that falls away from the cul-de-sac and runs next to someone's yard. The neighbor comes out and Underwood asks if Heather had let him know he was coming. Apparently not. "Damn Heather," he bitches, making a point to let the homeowner know of his disproval. I wonder if Underwood actually told Heather of our plans because this seems like a well-worn introduction. Regardless, the neighbor knows Underwood and takes no issue with us tromping in his backyard. We walk down the steep hill inspecting a series of ironstone step pools that accommodate the stormwater runoff from the neighborhood. These step pools have been constructed over the buried stormwater pipe that has, as a result of this project, been rendered obsolete.

The stone and cobble weirs that form the backbone of the conveyance system have gone through various name changes. First, they were known as regenerative stormwater conveyance (RSC) structures. Permit reviewers seized on the word stormwater and pointed to it as evidence that Underwood was using streams as stormwater detention ponds. Underwood and others morphed the preferred term into regenerative stream conveyance, which allowed everyone to keep the acronym but led to most people using two names for the same thing. Having been burned one too many times, Underwood now speaks only about the process (figure 2.2).

I was at this site approximately fifteen years ago—not long after it had been planted. At the time, it was a revelation. The berms of sand stuck out into the channel and formed sinuous oxbows pools along the valley wall. Unlike those formed along the Mississippi over decades of erosion and movement of the channel, these oxbows short-circuited time and formed a premeditated ecology. Planted on the sand berms were Atlantic white cedar saplings. At their base were small and inconspicuous carnivorous roundleaf sundews, *Drosera rotundifolia*, a delicate flourish of the planting palate. Few in the contracting community would select a plant of such diminutive stature. The survival rate is low, and its thin petioles add little to the requirements that 90 percent of restoration projects be covered with vegetation.

As we walk into the Howard's Branch project, the first thing I notice is the size of the trees. The trees are now twenty-five-feet tall in places and have filled

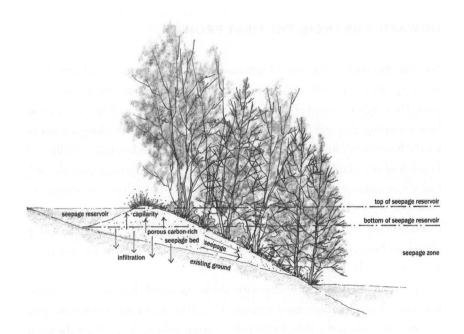

top of seepage reservoir

bottom of seepage reservoir

seepage reservoir   capilarity

porous carbon-rich
seepage bed   seepage

infiltration   existing ground

seepage zone

2.2 Drawing of the pool structure of a regenerative stormwater conveyance.
Courtesy of Brian McAveny.

in to form thickets. The density makes passage difficult, and in many places we are met with fallen trees that have blocked the path. Underwood bounces around them, finds a new route, and eventually stops at a point where water is seeping down from the hillside. "There are over 1,000 types of sphagnum," he tells me. "We seeded that area over there a couple of years ago. You can see where they are coming up." Indeed, a lush mat of sphagnum is heralding our arrival with a radiant green that stands out against the fall colors of the upland forest in the background. We continue to walk along sandy trails that intentionally follow the ridges of the sand berms. Sand berms make for low maintenance walking trails because weedy vegetation won't grow on them. Underwood points out a Mt. Laurel. "You don't see many Mt. Laurel's thrive around here. They thrive in sterile soils. Most people plant them in topsoil. If you want every invasive plant known to man, use topsoil in your project."

After Underwood recalls some facts about different species of sphagnum, I ask if there has been more attention to Atlantic white cedar these days. He mentions a recent publicity push by the state Nature Conservancy office and the National Aquarium from Baltimore to highlight Atlantic white cedar restoration. These two well-heeled organizations published several articles and highlighted some tree planting efforts at existing nearby preserves. No one reached out to him for advice or acknowledgment. The offense he takes seems to stem as much from the rather pedestrian approach of these projects as from the lack of acknowledgment. Planting trees in a preserve that once had Atlantic white cedar is a no brainer. Creating the precise nutrient and hydrological conditions in a small area jammed within the unclaimed spaces of suburbia is another thing altogether.

I'm curious about where this obsession started. Although it's not hard to imagine Underwood in his younger years, I had never asked before. Underwood says his family moved to a place close to the Severn River when he was thirteen. They bought him an inexpensive aluminum johnboat and "the world opened up." Starting from the day school let out for the summer until the first day of school in the fall, his days were occupied with fishing and crabbing in the bay. It was a time before fish and crab limits, and the bay provided a bounty to anyone who spent the time seeking it out. He says that aside from a short stint in California, he's never really lived anywhere else. "My parents could never have moved—they would have had to drag me away from this place."

I ask if there is more money these days for these types of projects. "The bankers from New Jersey are coming down to take the projects . . . and help meet mitigation requirements at a low-bid cost." Showing an uncharacteristic reserve, he holds his thought and stares at the sterile soil underneath. "You know, the thing is, this is ecological restoration—*this was supposed to be fun!*" Despite the apparent challenges, Underwood's projects have gained significant accolades. The grande dame of Atlantic white cedars and former Yale lecturer, Aimlee Laderman, told Underwood that she's never seen a more ambitious restoration project. His work has received more than a dozen awards, ranging from the "Best BMP in the Bay" to an "outstanding stewardship award" to an "Exemplary Ecosystem Initiative." Clearly Underwood has his fans.

KP and Chris join us as we cross a creek and continue walking along the other side. We come to a patch of bright green and short bamboo. "Arundinnaria—the only native bamboo," Underwood says. "I planted it over there five years ago, and they migrated over here. Guess they liked it better here." We continue along the creek to the end of the project, at which point the stream drops down a couple of feet over a jumble of boulders. I ask Underwood if this restored ecosystem was likely to have been here at some point in the past. He describes the accounts of Captain John Smith who plied the inlets and coves of Chesapeake Bay in 1608. "At five feet above saltwater, it was all pitcher plant savannahs. A psychedelic tapestry of reds, oranges, and greens of sphagnum and pitcher plants. It's hard for people to believe what they haven't seen in their lifetime. But that's what I go by."

It's an impressionistic and compelling picture, and it is one that I want to believe. However, it's unlikely that Captain Smith was focused on pitcher plants when he was stuck waist deep in a bog before being taken hostage by the local Powhatan Tribe. He miraculously escaped execution and later became the leading writer about the Virginia colonies. Captain Smith knew how to engage an audience and tell a good story. The veracity of his claims, bickered over by historians for decades, is honestly less interesting that the adventures he describes.

Although today's landscape would not be recognized by Captain Smith, he would understand the acrimony that seems to surround Underwood. Smith also wore controversy well. He was called into question by his peers and nearly executed for a series of unfortunate events that led to the death of all of a small exploration party he led. And similar to many of Smith's accounts, many of Underwood's claims cannot be easily verified. Smith's readers in England wanted compelling stories of the new colonies. Many of Underwood's followers want a story of renewal and restoration. A compelling picture told with a preacher's conviction is worth a thousand words in a peer reviewed science journal.

We sit at the outfall of this developing bog, watching the cedar stained "blackwater" flow down to the bay. To some degree, plant and stream bacteria have already begun their biological fermentation of these waters. In the nineteenth century, these stained waters were thought to precipitate contaminants and prevent the water from going bad. Captain Smith would have been well served if he filled his casks here. Underwood mentions that a population of yellow perch were found above this small grade change in the stream. "Not supposed to be there, so they say. 'We'll see,' I said. . . . Last year they found fry which shows that they have been reproducing. The researchers have them in formaldehyde in a lab somewhere."

"Can we put you in formaldehyde Keith?" Chris probes. This spurs a discussion of Underwood's future burial. Underwood is sixty-five, so this discussion is certainly premature, but apparently Underwood has given it some thought. He lets us all know he'd like to be wrapped in newspaper and pushed head down into the neighboring Alden Bog. "Just stand on my feet and push me down there!"

We head back toward the place from which we entered. Along the way Underwood nags KP to keep up and stay with the program. It's unclear what KP is supposed to be doing, but it must involve paying close attention to Underwood. I likewise stray from the program and make a wrong turn at one point. I take some pleasure in my confusion because it's uncommon to get turned around in a two-acre restoration site. The wildness of the place seems to be a positive sign of health for this not new, yet not fully established, ecosystem.

The transition from an ecologically restored site to a restored ecology has no specific criteria nor an established time line. As we bushwhack through the cedars and shrubs and sidestep patches of sphagnum, it feels as though we are somewhere closer to the natural condition. At the base of the hill, Underwood recoils and makes an unintelligible sound. His back is fighting back after forty years of physical labor by pinching a nerve. He shakes it off and continues trudging up the hill.

APPEAL AND CONTROVERSY

It's difficult to pinpoint the exact beginning and the precise source of the controversy. Controversy forms a backdrop to most discussions about stream restoration, and Underwood finds himself at the center of much of it. The 2005 Emily Berhnardt and Margaret Palmer survey of all stream-restoration projects nationwide received a lot of attention.[6] It warned of a widespread misuse of funds and a lack of proper monitoring to assess whether or not the project goals had been attained—if they ever had any stated goals. Another source stemmed from the orthodoxy that most reviewers and practitioners knew and supported. By 2005, most stream-restoration practitioners had been trained in a system developed and taught by Dave Rosgen. Most agency permit reviewers had also taken one or two of his classes. The fact that so many people had shared identical training meant that the script was clear and new approaches were not desired.

Underwood came to the game with an entirely different approach. Rather than assessing the desired channel type based on reference reaches, bankfull discharges, and position on the landscape, he saw an opportunity in these forgotten and degraded streams and came up with a solution specifically suited for these conditions. These deeply incised headwater gorges that had little to no biological activity could be used for slowing and cleaning stormwater runoff. His solution was simple but completely unorthodox. Underwood did this by filling in the incised channel, in some places twenty feet deep, with a mixture of sand and wood chips. This porous mixture ensured that stormwater would rapidly infiltrate into the engineered soil below. As it coursed through the media, it would be slowed, filtered, and reemerge in pools or riffles further downstream. To prevent this wedge of sand from eroding, step pools were installed every five to ten feet on top of the sand. These step pool structures would direct the volume from larger storm events to cobble weirs that served as controlled discharge points from the pools. The pools themselves would retain water for some period of time; the area would be planted with a mix of vegetation heavily sampled from the Atlantic white cedar species playlist; and a supporting cast of frogs, waterfowl, and other wildlife would appear later.

Important logistical factors contributed to support for these early projects. The first related to constructability. Nearly all of these locations were in areas where the slopes were steep and access was difficult. A typical Rosgen restoration approach would require the removal of at least a fifteen-foot-wide swath of trees on either side of the stream. Significant grading of those banks would be required as well, which would expand the footprint of the restoration project even further in steep areas. Standard postconstruction photos of stream-restoration projects show a forty-foot-wide graded area, secured with blankets of erosion-control fabric and planted with a bevy of new plugs, saplings, and small trees. The entire riparian area clearly has been "reset" to make way for the new, desired channel. Some studies have found that it may take decades for the vegetation to regain the height and structure to shade the stream as it once had done.

Underwood realized that he could work within the channel, essentially turning the channel into a haul road from which dump trucks could deliver their sandy loads. Once the channel was filled in, Underwood could use his front-end bucket loader to grade and construct the weirs. This immense movement of sand and stone was done with little or no footprint outside of the stream channel. In urban and suburban areas where most residents don't pay attention until a tree is cut down, this low impact approach had considerable appeal. Projects could now be completed without months of arguing with a parks department or a difficult neighbor about whether a design could be modified to save a particular tree.

The second major draw was related to the ability to capture stormwater runoff in already developed areas. Much energy and significant funding has gone toward trying to turn back our hydrologic impacts to our urban and suburban watersheds. Many high-quality ambitious projects have been completed. However, the nature of these types of projects, consisting of rain barrels, rain gardens, and other small-scale infiltration practices, makes that goal elusive. First, there is no way to force any private property owner to install these practices. Even the best social marketing campaigns will only result in 30 percent of the public adopting these practices. Given that some impervious surfaces can't be retrofitted, it would be necessary to convince more than 80 percent of private property owners to adopt these practices. Second, truly effective capture of stormwater runoff that can remediate our impervious sins requires a fetishistic obsession with capturing nearly all runoff from a property. Installing a rain barrel doesn't cut it. Even though many people want to do the right thing, few want to sign up for this level of commitment, especially if it involves significant cost to them or functional alterations to their yard. The RSC approach had the potential to meaningfully capture stormwater from a scale that hadn't previously been achievable. To capture 80 percent of the runoff from a twenty-acre subdivision is a major accomplishment for those in the trenches of this work. To do so with one project is undeniably compelling.

The third major draw was the allure of the habitat that could be created. This habitat was not generic. It was characterized by nutrient poor soils that were irregularly inundated with water and then dried out. These conditions had the potential to bring along certain ecological functions as well. On the surface, wildlife could take advantage of the ponded water. A more robust plant community could be established. Hidden from sight, water exchange between surface water and groundwater could be facilitated in the hyporheic zone. Denitrification could be enhanced in the saturated areas underneath and adjacent to the step pools. The steady mix of pollutants running off streets and parking lots could be trapped in the sandy matrix, a truly regenerative project. Whether or not this was the type of habitat that had been there in the past was secondary to the regenerative future: debatable but perhaps unknowable. Underwood knew it was the perfect habitat for Atlantic white cedar and the plants that thrived in those bogs. This threatened ecosystem, comprising less than 5 percent of the acreage of U.S. golf courses, could be established in these formerly ignored and biologically barren areas. Even if you doubted some of the promised functions, it would be difficult to argue that this new habitat was worse than what had existed before the project.

Not everyone was on board. One source of the controversy stemmed from the fact that these urban and suburban headwater streams, with their highly degraded habitat and low likelihood of improving, were still categorized as streams. To a reviewer who might be assessing a plan 150 miles away in Philadelphia, these projects were burying streams and effectively erasing the blue lines on the authoritative United States Geological Survey (USGS) maps. The step pools created didn't align with the approved principles outlined by Dave Rosgen. The Rosgen principle of creating a dynamic equilibrium whereby the channel could pass the median-sized particle that entered that reach was not followed. Rosgen's idea was to understand this sediment supply and create a channel that would neither aggrade nor degrade. Channel morphology was key to finding this balance. The step pools of the RSC projects, on the other hand, would clearly hold sediment and eventually fill up, according to some reviewers and other Rosgen-trained practitioners. There were concerns that the deep sand lens under these pools would also clog over time with the addition of finer sediment particles or, worse yet, the entire project would blow out and create a highly unstable mass of sediment that would smother downstream habitat. Whatever ecological functions these projects promised, it was thought unlikely that they could sustain these functions over time. How could someone approve a project when the future trajectory was so uncertain?

All orthodoxy contains blind spots, and this orthodoxy was no different. In urban headwater streams fed by stormwater, there is no significant sediment supply because the land has been supplanted by asphalt and rivulets by stormwater pipes. This reality fundamentally contradicts the academic understanding of stream function and sediment transport that stream restoration was based upon. No one had previously considered this because these degraded stormwater-dominated streams were generally not considered worthy of study until recently.

Another major source of controversy was visual in nature, and it set off alarm bells for many. Not long after completion, several of Underwood's projects were marked by bands of iron red flocculate that clung to stream boulders like snot from a toddler's nose. Although not uncommon in sluggish streams in the Coastal Plain, the flocculate was extensive in places and unsightly. The product of iron-oxidizing bacteria that thrive in conditions with low dissolved oxygen, a source of carbon, and some level of soluble iron, this flocculate attached to cobbles, stream banks, and logs. The flocculate could be agitated, and large stream flows flushed much of it downstream. No one could pinpoint exactly what caused this flocculate to occur, nor could people predict its impact. In some projects, the flocculate was so extensive that the stream streaked red throughout its entirety, reminiscent of a stream suffering the toxic effects of iron sulfide mining. Numerous theories were floated. RSCs raised the groundwater, creating conditions in which the existing soils were alternately saturated and dried. This created the conditions for extant soils and perhaps the sand fill to release oxidized iron. Bacteria feeding off this oxidized iron flourished. Others pointed to the iron-rich limonite boulders, an easy target because its reddish-brown streaks matched the color of the flocculate.

To the rescue came the University of Maryland researchers Solange Filoso and Michael Williams. They assessed this issue in a study published in 2016. Like many academic studies, this one did not entirely resolve all the questions, but it seemed to calm the fray. They found that the cobble and soil mix used in the RSC construction increased iron concentrations in stream water. Organic matter such as logs and wood chips added to the available organic matter. In addition, they recognized that RSCs act to slow the flow of the stream and create a slack water environment. Iron oxidizing bacteria thrive in these conditions and thus created the flocculate found in many of these projects. At the same time, these bacteria are not exclusive to these projects and are also able to use iron in stream water that may not originate from the added material but rather upland soils from which the stream drains. Perhaps most critically, they stated that iron flocculate "does not have a large impact upon stream habitat, but localized dense mats sometimes occur that embed habitat." Providing some guidance, they concluded, "considering the quality and strategic use of Fe-rich substrates and prevalence of organic matter in RSCs will likely decrease the incidence of flocculate."[7] In language the authors would never use, the study's findings could be summarized as "don't worry, not a big deal."

Another not so small regulatory issue was Section 404 of the Clean Water Act. This section prohibits filling wetlands or streams without compensatory mitigation. Although the Army Corps of Engineers is the lead in this review, the federal Environmental Protection Agency (EPA) and state agencies also have review authority in these cases. Section 404 impacts are anathema to most biologists in government agencies, and to create this impact in the name of restoration simply raised the ire of many of these reviewers.

Things got heated quickly. EPA prides itself on being neutral in relation to private businesses. Most realize that for their authority to be respected they must stick to the key mandates of their regulations and enforce them fairly and without excess baggage. Although specific businesses may be issued notifications of violation, each project or review is to be looked at with clear eyes and limited baggage. In Region 3 of the EPA, a Philly-style bureaucratic brawl had begun. Several biologists who were acting as reviewers were publicly accusing Underwood of illegal business practices. One reviewer pointed to the Palmer survey paper to prove that RSCs didn't work and cherry-picked other research to delegitimize the approach. Had alcohol been added to the mix, the situation might well have descended into the lawlessness of a Philadelphia Eagles game.

Tasked with sorting out this food fight was Nick DiPasquale, director of the Chesapeake Bay Program. His office, a high-profile separate unit of the EPA, was tasked with supporting implementation of the Chesapeake Bay total maximum daily load (TMDL). This job involved holding five states and the District of Columbia to account for massive sediment, nutrient, and phosphorus reductions and doing so on an unwavering deadline. It was the proverbial carrot and stick job because the Chesapeake Bay Program also supplied significant grants to these states to help them achieve these reductions. Over the period of several months, the RSC controversy rose up through the bureaucracies to land on DiPasquale's desk.

On one side were states and counties that said they needed these projects to meet their reduction goals. Not only were they effective, they were appreciated by the public everywhere they had been installed. Full-throated public support for projects is not the norm. DiPasquale appreciated Underwood's approach, but his support wasn't unlimited. "Keith has been practicing adaptive management before anyone was talking about adaptive management," he relayed, understanding that adaptation doesn't always fit well with regulation. On the other side, he had reviewers within his own agency referencing a section of an act fundamental to EPA's oversight of water. And to the letter of the law, what Underwood was doing was a violation of the Clean Water Act.

Resolution of these types of issues is typically not clear cut. Power is decentralized in most federal agencies. Workshops were organized to educate people but mainly served as venues for camps to assemble and eye each other suspiciously. Midlevel bureaucratic weight was thrown around. Ever the consummate professional, DiPasquale never got sucked into the mud. He worked closely with his counterpart at the Corps and created a special permit category for these practices. They would be exempt from Corps and EPA review as long as they were less than 1,500 feet in length. Guidance was drawn up to identify the best locations for these particular projects. Although the solution was more a negotiated truce than a decision, it seemed to lessen the heat by putting forth some guiderails that consultants and implementers can follow. It was a default victory for Underwood, but the ire of those reviewers did not melt away. Under the weight of this controversy, the issue compressed into a sort of sedimentary rock, crumbling off bits of invective when poked too hard.

BEER-THIRTY: TIME TRAVEL TO ARDEN BOG

It's midafternoon as we drive to the next site. The back roads of Anne Arundel County have seemingly never met the grid system. The roads wind around hills and valleys, bending to accommodate unseen inlets and coves of the Chesapeake Bay. Colonial buildings from the middle of the eighteenth century mix with larger new homes and roadside convenience stores. It's a landscape that you need to know well to navigate.

History infiltrates the area as well. Some 238 years earlier, General George Washington and his French pal Rochambeau made a historic trip down the same road we are currently traveling. Fully aware of his need for allies, he employed the assistance of the French major general whose troops would help deliver the critical victory over the British at Yorktown and provide a path to independence. As the two cut through this area, they likely avoided the bogs in favor of dry land.

I try to get Underwood's perception on the bad blood. "What's behind it all," I ask. "Is it just plain competition?" He returns to the "New Jersey bankers" who are swooping in to do work on the cheap. They deliver the credit the municipality needs in an apparently mercenary manner. I ask him about other firms he has worked with in the past that have a restoration focus. "They got what they needed out of me, so I don't hear much from them anymore." I think to myself that Underwood could use a Rochambeau right now. We stop to check in with the other cars in the caravan and Underwood goes in for a pit stop. Three minutes later he returns with a twelve-pack of Heineken and stashes is at his feet in the passenger seat. Without asking, he pops one open as we head out to the next site.

I had asked to go to the best-preserved cedar bog in the area. As I'm driving and Underwood is drinking, he talks about how Arden Bog is the best but is not quite pristine (figure 2.3). A new enemy has appeared in the past several years and wiped out nearly all of the pitcher plants. "When I first saw the place, it brought

2.3 Keith Underwood points out vegetation bands at Arden Bog.

me to my knees! Now the beavers have damned it up and flooded out the pitcher plants." Underwood naturally has a plan for it and is working to convince people. It involves taking up an entire road that cuts through the valley and serves as the berm supporting the marsh. He envisions a porous stone base that will make any beaver activity irrelevant because the stone lens will keep the water at the proper level. Something tells me that the state highway administration may take issue with water passing under their roadbed.

Underwood mentions another person involved in a bog restoration project—a member of the restoration diaspora. This person is working on bog restoration in Virginia and had once borrowed some pitcher plants as stock for his project. With those plants now forming a healthy population, Underwood had asked for some specimens to restock Arden. He was turned down. "I'm a plant person, but plant people are assholes," he confides. As a plant person myself, I understand his point. Obsession about this immense and diverse kingdom of life can lead to some antisocial behavior. It's not that people are bad but rather that plants are frequently just more interesting.

As we approach Arden Bog, Underwood points out the landscape. We're on a sandy ridge that pitches slightly away from the valley. Adjacent to the edge is county parkland, dotted with some basketball courts and picnic areas but mostly lacking the impervious cover present in other areas. Underwood points to this as the reason the bog has remained in its high-quality state. We take a right and start down from the ridge. After a few minutes we pull off the road and get out to survey the area. Stretching out in front of us is a saturated landscape covered with mats of sphagnum moss, rusty colored blueberry shrubs, and other submerged aquatic vegetation. The bog is notable for its incongruent remoteness. There are no signs of houses, cell towers, power lines, or other visual detritus of our modern landscape. We have descended one hundred feet down into a valley from suburban Maryland and landed in Nova Scotia or northern Wisconsin. The bog is surrounded by a band of leather leaf shrubs, followed by magnolias and holly trees that claim the solid ground up to the base of the hill. Virginia pines, maples, and oaks fill in the valley hillside, completing the protective buffer around this ecological gem.

Underwood is explaining this transition of vegetation to me as I take in the view. He points to the bands of vegetation clearly visible from our vantage point. He mentions an unwanted species, spatterdock, which he says releases marl that elevates the pH of the water, making it less suitable as bog habitat. He explains that this ecosystem can only thrive in nutrient poor, acidic, and saturated environments, and the key element that prevents excess nutrients here is the semipreserved landscape and associated hydrology. All rain and related runoff infiltrates through the sandy soil rather than coursing overland. Any fertilizer runoff from the surrounding yards is either bound up to the sand particles or taken up by plants. "Seepage good, pipes bad," he says, realizing that this complex hydrology and interdependent mix of species could use a bumper sticker summary. Despite this streamlined summary, it is clear that Underwood has absorbed every facet of this particular ecosystem. These relationships and processes form the tool kit he uses in his projects. His particular ingenuity is re-creating these processes within the limited confines of space and permission in which he navigates.

The tranquility is interrupted when KP takes a call on his cell phone. He's quietly trying to schedule an inspection of a site within the fourteen-day window listed on the permit. He's trying to set the date five days from now, apparently to close out the project. The person on the other end of the line is trying to push it back to the end of the fourteen-day period. "Five days!" Underwood blurts out. "They don't have to be there! Tell them five days!!!"

FUSSY PLANTS AND UNCERTAIN PROFITS:
THE FUTURE OF RSCs

In the business of sculpting and restoring nature, no patents are awarded by mother nature. Ideas are regularly stolen and incorporated in other projects. Any trade secret is no longer a secret once the project has been completed and is in full view of all. Underwood's projects have been racking up awards in the mid-Atlantic, and other stream-restoration practitioners have been adopting and pitching the approach in far flung areas. As far away as Milwaukee, Wisconsin, firms have installed a more muscular version of the RSC for similar problem areas. Many development projects faced with costly, space-demanding stormwater control ponds are closely looking at whether they could install an RSC and potentially save space and money in the process.

The approach has been formalized in a manual by the Maryland Department of Natural Resources. Other practitioners have led trainings on the approach, hoping to systematize the approach and avoid failures in the process. Underwood's ecological alchemy is now a black-and-white line drawing that can be inserted into a plan. Despite the good intentions of this standardization, I can't help feeling that something is being lost.

I wonder if the RSC approach will be used like a mercenary army to meet increasing permit requirements at lower cost. What is the motivation to get the soil or hydrology just right when specialized plants are substituted with generalists? How many backhoe operators have an aesthetic bent when placing boulders? Does it matter? Where is the incentive to create threatened habitats when the permits only require a certain "vegetative cover." What will ensure that these practices add up to more than a giant sand filter dropped into a stream channel? Where is the profit motive in obsessively tending to fussy plants and in the process creating something unique and beautiful?

SAINT LUKE'S: SALVATION IN ANNAPOLIS, MARYLAND

I told Underwood that I wanted to see one of his more recent projects. I wanted to see how the Atlantic white cedar restoration and stormwater control projects had evolved. He said he had just the place—a project that brought everything together. Fittingly, he took me to a church.

The St. Luke's Episcopal Church project in Annapolis, Maryland, is located within the urban mix of Annapolis. Because of the trees, houses, and a hilly terrain, you would never know it is less than a thousand feet from a cove of the bay. The bay seems to extend its watery fingers into even the most agnostic commercial sprawl of the city. The church itself is a simple brick building surrounded by a long parking lot on one side and an open grassy area on the other. A labyrinth is situated in the grassy area, and I think I might need to walk it to sort out the endless stream of facts and claims Underwood sends my way. He walks me down the parking lot to a point where a very low footbridge spans a cut in the curb. It's an innocuous start of the project, more puzzling than impressive. He begins with the creation story.

This project was initiated by Betsy Love, a graduate of the Watershed Stewards Academy, a free watershed training program that requires a capstone project that usually involves construction of a rain garden or installation of a couple of rain barrels. The student told the teachers that she wanted to do *this* project and she wanted Underwood to build it. She then helped get a grant from the Maryland Department of Natural Resources to make it happen. Great tidings come to those who believe.

Underwood happily relates this request because he always prefers to be the one operating the heavy machinery. We're standing around in the parking lot, and Underwood is sharing how the government inspectors once demanded that he personally be replaced as the operator of the bucket loader on a particular job. The inspectors took issue with his rapid and efficient placement of stones. KP replaced him, and Underwood is presently acting out a painfully deliberate placement of boulders. This pantomime is uncomfortable because it appears that he is ridiculing his staff, but the punch line is still to come. At several follow-up meetings, KP received accolades from the inspectors on his expert bucket loader work. Guffaws erupt from the guys. KP had never operated that machine before and had been slow walking it due to inexperience rather than care. Ye shall make no false idols unless the inspectors tell you to do so.

We refocus on what is front of us. Underwood points out that we're at the top of the watershed, and this little footbridge was added to address a request to keep the sidewalk dry in this area. He points to the low damp area just past the bridge. "This is what we're calling the rain garden," he says. His previously staccato speech is becoming more legato as he walks me through his creation. "What we're doing is breaking up the watershed into its smallest constituent parts. It's what nature does on its own." Nature is a temple unto itself.

Next, we walk over to the sidewalk adjacent to the main two-lane road at the top of the watershed. We're standing on the catch basin that collects the runoff from approximately an acre of roadway. I sense that something is not standard about this particular catch basin when I see an artfully arranged jumble of ironstone under a new opening in the back of the catch basin. Underwood directs my attention to the interior of the concrete vault where the opening of a twelve-inch-diameter pipe is just visible. This pipe, he informs me, extends all the way down to the cove and transports runoff from both the adjacent street and other drainage areas further up in the watershed (figure 2.4).

This hidden pipe is part of the remnant conveyance infrastructure that is the conduit for nearly all of our urban pollution ills. They are ubiquitous anywhere Underwood works. Any other contractor would see the removal of this as perhaps a costly but necessary element of the project. Digging up stuff provides a lot of billable time. Underwood's genius stems from the fact that he doesn't see this as just another line item in a project but rather as an opportunity to innovate. His process, which he calls a "bubbler," involves filling this pipe with cobble and gravel so its purpose is transformed from a conveyance structure to a storage structure. This effectively changes the source of the problem (uncontrolled and concentrated flows) into a solution (stored water that is released slowly). Underwood anticipates that the pipe will plug up over the course of several months or years when the sand and other fine material work their way down and clog the pipe. Process trumps exact time lines in this service.

This pipe then bubbles up the now slowed and tamed stormwater, pushing it out the new opening of the catch basin and into the top pool of his conveyance structure. This approach doesn't exist in any spec sheet. This minor tweak is remarkable for the fact that no other contractor or designer would ever think of, let alone be comfortable with, the do it yourself repurposing of a now unnecessary pipe. The client has saved a significant amount of money, and Underwood has subverted existing infrastructure for his purposes. Wealth gained by fraud or bloated line items dwindles.

Underwood does not trust labels because those he has come up with tend to be thrown back in his face. His terminology runs loose, but his understanding of the process is complete. As we walk through the site, he says "we're calling this the rain garden and transitioning around the back to a grassy swale." Seeming to have regained clarity, the grassy swale runs by the labyrinth and drains into small ironstone-lined step pools planted with a few Atlantic white cedar saplings. The

(A)

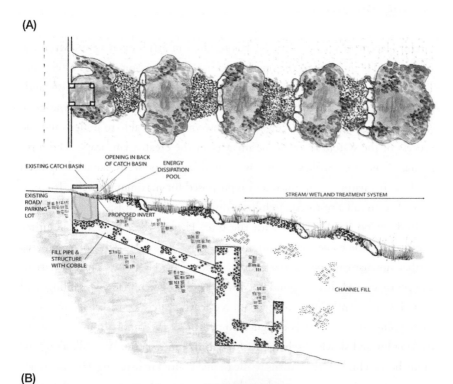

EXISTING CATCH BASIN

OPENING IN BACK
OF CATCH BASIN

ENERGY
DISSIPATION
POOL

EXISTING
ROAD/
PARKING
LOT

PROPOSED INVERT

STREAM/ WETLAND TREATMENT SYSTEM

FILL PIPE &
STRUCTURE
WITH COBBLE

CHANNEL FILL

(B)

2.4 (*A*) Drawing of the "bubbler" retrofit for the catch basin and drainage pipe with the RSC constructed and planted above. (*B*) Keith Underwood and Keith Pivonski at the retrofitted catch basin near the St. Lukes RSC project. Courtesy of Brain McAveny.

careful placement of the ironstone has resulted in artful openings between the structures. He motions to the opening to see if his creation invokes a reaction. The landscape triggers fond memories of bouldering through the canyon lands in Utah or weaving through sea stacks on the Oregon coast. The gaps in the stone create a recognizable visual language of movement. I'm able to understand the process of the system through the imagery of the created landscape. The rusty patina of the stone makes the newly created landscape seem like it's long been in place. It's nature reimagined and repurposed for maximum benefit. It seems farfetched to call a restored ditch beautiful, but somehow it is.

Underwood isn't finished with the tour. He has the energy of a minister approaching the come to Jesus moment of the sermon. He calls me to an area further downstream below the church parking lot. He points out the narrow and gently curved shallow pond lined with verdant green magnolia saplings and dotted with recently fallen orange and red leaves. "Here's our seepage reservoir, we're going to call it." It is designed to detain all the runoff from this sizable parking lot and slowly discharge it through its sandy berm. We walk along the sandy berm a bit before he turns to me to make sure I'm listening. His insistence peaks as he points out the pitcher plants growing only six feet away down the sandy slope (figure 2.5). "This pitcher plant, one of the pickiest plants in terms of water quality needs, grows only in a *pristine habitat*. This dirty stormwater is directed into this seepage wetland and is now feeding these plants." The fact that this finicky plant appears to be thriving is proof positive that he's right. This charismatic plant is normally found in nutrient poor bogs that have developed over hundreds of years in northern hinterlands. Underwood has created a happy home for them over the course of a nine-month project. The reward of this faith is to see manifested what one believes.

THE RECKONING

Fifteen years after first seeing Underwood's approach to restoration, I still yearn for clarity. The potent mix of swagger, conviction, and iconoclasm is appealing to me. It's a perspective that declares that half measures are not enough, and that to accomplish something worthwhile, you may have to ruffle some feathers. But I want clarity on how effective this practice can be at rectifying numerous impacts to our streams and rivers. Hydrological conventional wisdom tells us

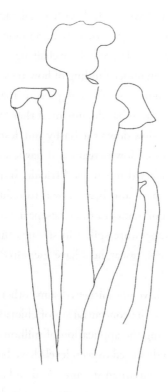

2.5 Trumpet pitcher plant.
Drawing by Claramae Hill.

that the game is basically up once you hit 10 percent impervious coverage. Macro-invertebrate and fish communities simplify and degrade. Our suburban areas range from 15 to 30 percent and our urban areas 30 to 50 percent. If this approach stands up over time, it would exceed the benefits of nearly all completed restoration projects. If not, it may be another well-argued point on a misguided track record of half measures. How many more chances do we have to get it only partially right? It's a question without an answer, but I'd have to guess not many.

Many researchers are trying to figure this out. But it is not simple, and the results are not definitive. If understanding natural systems is complex, understanding anthropogenic impacts on natural systems can be particularly knotty. Teasing out the influence of restoration approaches in these human impacted areas is especially involved. Researchers working in wilderness areas may have to contend with three-day hikes or hungry bears, but researchers trying to understand these impacts have to deal with small sample sets, projects that evolve over time, restoration projects that are never exact replicates, and the lack of truly comparable study sites. Despite these challenges, information is trickling in and beginning to form an understanding of how these projects function.

The results paint a more complicated picture than Underwood's vision. Supporting his vision, research has documented that Underwood's restoration projects have an ability to slow down the flashy and destructive pulses of stormwater. This prevents erosion downstream and havoc on a range of infrastructure. The cost of those averted repairs is difficult to quantify but is certainly significant. The projects have also been shown to reduce sediment and nutrient concentrations, key pollutants in the Chesapeake Bay. They have also been shown to reduce the temperature spikes in streams that follow summer rainstorms. In unrestored areas, researchers have measured five-degree spikes that essentially scald aquatic life.

Underwood's projects have not mitigated some other impacts. They have not shown an ability to reduce the concentration of chlorides that spike in the late winter and spring following the application of millions of pounds of road salt. Likewise, no increases in dissolved oxygen levels have been measured, a limiting factor in most urban and suburban streams. And so far no studies have shown a rebound of pollution sensitive aquatic macroinvertebrates that are both key indicators of stream health and the building blocks of the aquatic community. As with all research, the final word has not been written yet.

In the face of these inexorable development trends, we have reason to be hopeful. Public support for Chesapeake Bay restoration is strong regardless of political affiliation. There are signs that things are improving. Because of unprecedented regulatory actions mandating restoration benchmarks on a range of stakeholders, efforts are more robust than anywhere else in the country. The scorecard put out by the EPA in 2018 shows a 105,000-acre increase in bay grasses, and 42 percent of the bay is meeting water quality standards. Can we translate this interest in our well-known ecological assets to backyard stormwater gullies? Can these places become ecological outposts, sustaining a complex and interdependent ecosystem in unlikely spots behind our backyard trampolines and leaf piles? As Underwood would say, "We'll see."

CHAPTER 3

LEGACY SEDIMENT

Dorothy Merritts and Robert Walter Dig Back in Time in
Lancaster County, Pennsylvania

LITITZ RUN, LANCASTER COUNTY

As I walk toward the stream, the ground transitions from hardpan farm road to squishy damp wetland. The floodplain is saturated, thanks to a recent rain that had pushed the river beyond its banks. I walk through fifty feet of spongy sedge and grass wetland, verdant green from a rainy summer.

Tussock sedge forms dewy clumps, and the football-shaped spikelets of spikerush splay out gracefully from pendulous stems as if re-creating a fireworks display. Before I reach the main channel, I come across shallow rivulets that wander through the dark organic wetland soils. As I approach the main channel, the wetland yields to it without any of the stone armoring I was expecting. I need no monitoring equipment to know that this restored stream is healthy and vibrant. The water flows as clear as the cloud-free sky above.

Surrounding me is the bucolic landscape of Lancaster County with its rolling fields and picture-perfect barns. Driving down the roads, regular sightings of the horse and buggies of the Amish make time feel fluid and pliable. Lancaster County contains the car-dependent development that is seen almost everywhere else in the country, but the presence of the Amish somehow references the past even though their culture is very much alive in the present. This shifting of historical benchmarks plays into the restoration approach for the streams in this area.

My visit to this restoration site, Lititz Run, is eye opening. I spent seven years attempting to restore a stream in the District of Columbia. The stream, literally walled in, had been afflicted with the complete range of insults we unwittingly send their way: sewage main breaks, collapsing banks, and all manner of trash from shopping carts to tires. Restoring some sort of natural function to this water body was an uphill battle in the best of situations, and long-term success seemed unreachable.

Seeking some perspective, I made my way to this restoration site nine miles north of Lancaster, Pennsylvania. My friend and former colleague, Mike Rhanis, had been telling me about the research and stream-restoration work in which he had become involved. His former professors at Franklin and Marshall College, Dorothy Merritts and Robert Walter, were taking a different approach to restoring the streams in this area. "It is worth the trip," he told me.

In its most successful projects, ecological restoration has the power to return the function and beauty of our ecology in a way that is understandable on a visceral level. The part of our brain that has evolved with nature realizes that the web of biota lurking in the plants, soil, and water can sustain us. Call it an ecological baptism or just an impressive site visit, the rightness of this place was self-evident.

Eight years later that trip to Lititz Run is still as fresh in my mind as it was that day. It demonstrates that our society has the potential to fully repair our collective insults to our water bodies. The project aimed not just to stabilize the impacts but to return the countless and interdependent ecological networks to full health. Although I needed to find out more, it pointed toward that promise of restoration. Sometimes called ecological uplift or a regenerative ecology, it is what those restoring rivers and streams seek. It's the ecological connectivity of the Round River that Leopold evokes in the *Sand County Almanac*. I decided to return to Lancaster to investigate these stream projects further and to take in its hopeful waters.

FRANKLIN AND MARSHALL COLLEGE AND WALTER AND MERRITTS'S LAB

As I drive through town to the campus of Franklin and Marshall College on a fine late October afternoon, I'm buoyed by a sense of optimism. The leaves still clinging to the trees have turned from an orange-red to a rusty brown. The campus strikes me as how a campus should feel. It radiates the sort of peace and tranquility most liberal arts colleges strive to achieve: safe, leafy, and coursing with an intangible feeling of aspiration. Old Main provides instant gravitas; its four stately gothic pillars occupy the center of the campus. It is surrounded by a mix of older brick buildings and more modern structures that bear the names of donors.

Trim federal-style row houses with ground level porches fill the neighborhoods surrounding the campus. I can imagine students congregating on these porches on weekend evenings, sharing a few Yuenglings while someone breaks out a guitar. Although it's only the afternoon, I feel an uncharacteristic urge to strike up a conversation with some strangers. Everything seems overwhelmingly comfortable.

I meet Dorothy Merritts and Robert Walter next to the physical sciences building on campus. Its Sunday and there aren't any students around, so they drive up the sidewalk to unload several things at the door. Merritts is wearing a red Stanford sweatshirt and welcomes me with a warm smile and an energetic handshake. Walter also welcomes me and sports a neat mustache that suggests a level of earnestness that eclipses hipster irony. As they let me into the building, they tell me their room number and forge ahead, mentioning that we can meet in Merritts's office in a few minutes.

Heading down the hallway, I am waylaid by the maps of tectonic plates, fault lines, and other geologic features that line every wall (figure 3.1). The maps have

3.1 Dorothy Merritts, Bob Walter, Mike Rhanis, and Logan Lewis in the hall of the physical sciences building on the campus of Franklin and Marshall College in Lancaster, Pennsylvania.

a level of familiarity in the broad outlines of continents but diverge to explain other subterranean worlds unfamiliar to me. Fault lines are depicted in unlikely locations. Wisconsin is calved by one, jeopardizing its bovine tranquility.

The offices of Walter and Merritts occupy a large swath of this floor in the geosciences building. Although the two share a full-time position, they have ample space in their individual offices, a computer lab, and in other rooms where they can conduct experiments. In their case, the family business is the vast sweep of geological processes rather than the wide range of home heating and cooling options. It's an arrangement that seems to work.

It's somewhat unlikely that they find themselves in this tranquil spot in southeastern Pennsylvania. Walter had come from a uniquely expansive position in the state department where he was part of a small team charged with assessing factors that contribute to complex humanitarian disasters. His team was tasked with identifying which social, environmental, or climatic factors might predict catastrophic conditions with the hope of getting a jump on the problem before it fully manifested itself. Prior to this he had done work in the Rift Valley, carbon dating ancient settlements and, most notably, "Lucy." For geologists, this is sexy stuff. Many years later, after meeting Merritts, he found himself back in Lancaster County where he grew up, studying the same streams he knew as a kid.

Merritts came to Lancaster from the West Coast where she was involved in early work dating ancient terraces. Advancements in technology have provided tools that now occupy her successors who are building on her analysis. She seems happy that others are carrying on her work. Raised in rural central Pennsylvania, her work has also taken her around the world. Her study of fault movements led her to Korea, Indonesia, and Australia, and the geologist's pick on the table attests to these field days. Recognitions and awards are plentiful in her curriculum vitae but perhaps most telling are the Outstanding Educator award from the Association of Women Geoscientists and the Kirk Bryan award given to her and Walter by the Geological Society of America. The latter is the standard bearer in the field, and it recognizes the best paper of the year. This was the first clear recognition that the two of them were onto something.

It's also unlikely that two geologists would find themselves immersed in the current discussions in stream restoration. The lens by which geologists view the planet peers back millions of years. This lens comes into clear focus where pressing issues are at hand: cataclysmic earthquakes, lucrative oil and gas deposits, and sinkholes that threaten the foundation of neighborhoods. The scope of their investigation is often vast: mountain ranges, aquifers, or deposits that span multiple states. It's much less common for geologists to weigh in on site-specific restoration approaches that seek to address long-term impacts at this precise moment. The questions raised might seem myopic or excessively extant.

Bucking this convention, Merritts and Walter's current work is drawing a larger audience these days. The term *legacy sediment* now pops up in far flung depositional areas, ranging from the Chesapeake Bay Executive Council to *Science* magazine. Despite their respective credentials, it is an anomaly that their work is finding this larger audience. Most influential scientific research comes from larger universities, not liberal arts colleges. In their current positions, Merritts and Walter aren't able to assemble a small army of graduate students to help expand their academic reach. Nor is it likely that they will secure the million-dollar research grants some university labs can access. Despite these limitations, they both seem content to operate out of their corner of the academic universe.

To advance their work, they rely on a network of academic colleagues and former students. They've pulled in an economist at the school to look at the economic aspects of their work. They've found some grant funding to keep two former students involved in their mapping work. Logan Lewis is working on the presentation for tomorrow; he graduated a year ago and is weighing options for graduate school. He seems content to forego a return to his rural Nebraska hometown of 200 people and remain in Walter and Merritts's orbit. Mike Rhanis, the other former student, has provided critical remote sensing and geographic information systems (GIS) analysis that has unlocked Walter and Merritts's discovery, and he is frequently listed as a coauthor in their papers. Preternaturally methodical, Rhanis continues to work on GIS analysis of these rural Pennsylvania streams. His multifaceted role of friend, academic admirer, and on-call analytical wizard seems to serve everyone well. This is a select club of legacy sediment geeks, and their primary mission is to get people to understand exactly what comprises the ground underneath them.

LEGACY SEDIMENT EXPLAINED

In the summer of 2003, Merritts and Walter found themselves wading a second-order stream in Lancaster County, helping a graduate student navigate some confusion at her field site. Their student was puzzled by the composition of the bank material. Geologists are trained to see and understand the movement of all earthly materials that have been blown, deposited, or erupted over time: lateral accretion, fining upward, the angle of soil layers, and the inclusion of various heterogeneous materials such as glacially worn stone or pebbles. The terminology can be uniquely enigmatic, but it helps describe the processes that create the geologic context of an area. The student had found a wall of homogenous silty material: fifteen feet of undifferentiated soil sat inscrutable in this Piedmont stream bank. It was a geological Rothko, challenging the viewer to make sense of it. Without any clear indicators, how could this stream bank reveal its story?

Drawing on her past work studying ancient terraces, Merritts and Walter quickly identified the source of these layers as pond sediments. Recalling his childhood in the area, Walter remembered the network of millponds he occasionally found during his fishing excursions. Now with greater purpose, the three wade downstream to a point where they found large, rough-hewn blocks of limestone. Walking up a more gradual bank at this point, they noticed that these blocks continued up the slope and eventually connected to a stone race. They were surveying the remnants of a fifteen-foot-high stone dam that once provided power to a local mill.

In hindsight their discovery seems self-evident. At the time, however, no one was thinking about these riverine areas in their historical context. People knew that mill dams were present in this part of the country. What people had not realized is that this marker of colonial industriousness, placed every three to four miles along most streams in the area, had a transformative impact on nearly every stream in south-central Pennsylvania. Few people understood that the original settlers of this area came with highly developed milling technology from the old world. "Water power was the oil and gas of today," Merritts notes. Nearly all aspects of early American life were dependent on these structures.

For the streams themselves, it changed everything. For our idea of what a stream should look like, it presented a challenge. Our image of a single stream channel winding through a rural landscape is based on this wholesale damming.

It is not an accident that this notion of a stream is exactly what most stream-restoration projects are trying to create. Reinforced by personal experience and artistic representation, this image may accurately represent the last one hundred years, but it is neither historically nor ecologically correct.

What you would have seen would not be recognizable today. You would have seen a wide, marshy floodplain filled with lush wetland vegetation. The stream would have been composed of several connected braided channels—an anastomosed network of numerous impermanent channels weaving through this marshy expanse. During rain events, the river would slowly rise to cover the entire floodplain, depositing sediments on the surface. The wetland vegetation would pull out phosphorus and nitrogen from these sediments, creating denser network of roots and increasing the stability of this soggy, supple landscape.

Over the next four years, they dug deeper. Rhanis located old millpond sites by tracking down historic maps and comparing them to current stream maps. They then went to these sites and excavated ten-foot-deep trenches in the floodplains to reach the original floodplain. They conducted freeze-thaw tests in the lab. Through this multipronged approach, the picture became clearer.

The results of the fieldwork were definitive. In nearly each location where historic mill dams were present, a three- to fifteen-foot-deep layer of uniform, unconsolidated, silty sediment was evident. This "wedge" of sediment continued upstream, sometimes for several miles, gradually becoming thinner as the slope of the valley tapered its depth. Underneath this thick wedge was evidence of earlier wetland soils: dark organic soils representing a seed bank from those past wetlands. Unearthed by massive stream bank collapses, this clear black line revealed itself as the historic elevation of the braided stream and wetland landscape.

The term for this process doesn't immediately convey the massive change in the landscape that it describes; *base-level change* describes the accretion of these sediments behind the damns. Unlike the dramatic burying of Pompeii, this deposition happened over the course of many decades. Concurrent with the settlement of the area that began in the mid-1700s, mill dams were constructed along the streams to meet the needs of these early colonists. The widespread clearing of forests ensured that huge amounts of sediment would travel down from these watersheds. This sediment would settle behind these mill dams in the still water, silently accreting just under the surface. Fast forward to the early 1900s, when stream stewards began to breach these outdated dams, either for safety concerns or due to structural failure. With the water level now lower, these wedges of sediment were exposed, and vegetation quickly began to grow. After the rapid growth of sycamores, elms, and other riparian vegetation, these banks would frame the streams that we recognize today. However, this current iteration of a stream carried with it a troubling legacy.

These deposits of sediment were not content to stay put. Their unconsolidated nature meant that they were highly susceptible to the freeze-thaw process that leads to iceberg-like calving of banks in the spring. These collapses of sediment into the stream channel mobilize tons of sediment with phosphorus attached that rapidly flows downstream. As the stream incised into these floodplain sediments, it found increased energy and began to rapidly erode the banks. In terms of the allowable budgets of sediment and nutrients for the Chesapeake Bay, these wedges of sediment represented a ticking time bomb. Awaiting near certain erosion, collapse, and transport, these legacy sediments were now seen as a wholly unwelcome historic anomaly (figure 3.2). According to Merritts and Walter, the time to deal with this legacy was now.

3.2 Mill dam on the Wessshickon [sic], 1905.
Library of Congress, https://www.loc.gov/item/2016804782/.

THEORY + DATA COLLECTION + PREPARATION = IMPACT AND STRAIGHT TALK

We convene in Merritts's office where Lewis is working on a presentation. Their preparation is honed for a consequential meeting that will attempt to bridge the gap between academics, regulators, practitioners, and advocates. Tomorrow they will be hosting a bevy of government staff from the Maryland Department of the Environment and their counterparts from Pennsylvania, as well as several people from watershed organizations and other nonprofit groups. They've invited them to Lancaster because the restoration sites are nearby, but also because Merritts and Walter can tag team the field trip and still teach their classes. The fact that we're here today, on a Sunday afternoon, doesn't seem to strike anyone as odd.

We're looking at maps of the streams of Lancaster and surrounding counties that Lewis has brought up on his computer. All are tributaries of the mighty Susquehanna River, a massive watershed for Chesapeake Bay. Its forested head-waters are in a small part of southern New York State, but it gains its agricultural character in the counties of central Pennsylvania. It completes its journey through Maryland, but before flowing into Chesapeake Bay it is interrupted by the massive Conowingo Dam. This dam is the elephant in the room for sediment dynamics in this watershed and is a major concern for Chesapeake Bay overall. By 2014, more than 174 million tons of sediment had been captured behind this dam, with an additional 4 million tons being added each subsequent year.[1] This bathtub is now nearly full from the ninety years of erosion and the 27,000 square miles draining into it. This, however, is only the most obvious player in the story of interrupted sediment.

We're straining to view Lewis's monitor while munching chocolates that Merritts brought. Rather than the usual blue lines of most maps, the streams are delineated by a patchwork of red polygons of varying size. Their choice of color is not an accident. They want everyone who sees these maps to feel some sense of alarm. These sites are the result of several years of research, fieldwork, and GIS analysis. Measuring their exact dimensions is of critical importance. On one slide Lewis has tallied the sum of these red polygons, and Merritts asks the group what they think about this summation. They discuss some of the technical underpinnings of the analysis. After listening to the thoughts of Rhanis and Lewis, Merritts says, "I really like the way you describe that Mike. I agree—we need to be conservative in our assumptions on this." I'm struck by the egalitarian nature of the discussion, expecting something more hierarchical and formal. In an easy manner, Walter and Merritts draw out the thoughts of these two former students, preparing the two for their presentation. For my part, I understand about half of what they are debating. But the process I witness makes me want to head over to the registrar and sign up for the next semester; it's stimulating, supportive, and dynamic.

This being an academic setting, there is discussion of other researchers. Merritts mentions Reds Wolman, the father of modern fluvial geomorphology. Conducting much of his work in nearby Maryland, Wolman developed the scientific principles of sediment transport in streams. His work described how sediment passed through streams and exactly what sized particles could be expected to move based on the dimensions of the channel and the valley. This movement was a dynamic process. Not all mobile sediment ended up in a down-stream lake or bay. Some of it remained in the stream, forming new depositional bars. Wolman described this entire process, and it remains the foundation for the collective understanding of how water and sediment interact in a stream channel. It's no accident that an invitation-only meeting of who's who in geo-morphology and stream research is affectionately called the "Wolman's Club."

Merritts and Walter are well versed in this foundational work, but Merritts is now describing geologic processes that expand past the time frame of chan-nel dynamics that were explained by Wolman. She excitedly describes a time they saw evidence of a Pleistocene boulder fan buried under legacy sediment. "Fist-sized boulders interspersed in otherwise organic dark wetland soils," she explains, with the energy of a basketball fan relaying a last second three-pointer for the win. These boulders were not placed there by a stream, she emphasizes, as if I was foolishly ready to credit everything to rivers. She explains how the gelifluction or sloughing of boulders down hillsides resulted in these remnant geological features. I find myself getting unexpectedly excited by the image of a slowly moving landslide of prehistoric mud and boulders.

Walter mentions accounts of early milling technology that they discovered as part of their research. "The proliferation of these mill dams was the grammar school for the Industrial Revolution," he explains. If these dams taught lessons that would be further exploited in the Industrial Revolution, Walter and Merritts are providing a postsecondary lesson aimed at unlearning this formative and problematic education.

Set within this 300-year social and economic evolution are the implied but less clearly stated goals of stream-restoration projects. All projects make some claim to be self-sustaining. They make these claims in various ways: from being "maintenance free" to "finding a dynamic equilibrium" to "resisting lateral and vertical adjustment." These terms all speak to the major question: "Will this project stay in place, or will someone have to come back in five years to fix it?" This question is rarely addressed head on for the primary reason that no one can guarantee stability and few are unwise enough to do so. Each project is a leap of professional faith nudged by a belief in hydraulic analysis and a sense that something must be done.

I ask them which specific stream projects have shown the greatest stability and for how long. This is a difficult question to answer given the relative youth of the field. People have been stabilizing banks for decades, but concerted efforts to design and sculpt stream channels with the goal of long-term sustainability only began in earnest in the early 2000s. Most of the projects have been completed in the past decade, so the jury is still out.

Merritts mentions that they have found evidence in the seed bank buried under ten feet of legacy sediment that the flora of these wetland streams was stable for five to ten thousand years. Stable vegetation is a very good indication of a stable channel. Not only do they make this bold scientific claim, but they imply that the warranty period for all stream-restoration projects accordingly should be expanded by 5,000 times what is currently promised.

Someone mentions a project completed several years ago in Maryland, just over the border. Lewis pulls up the aerial image on Google earth. A series of distinct white parabolas, evenly spaced along the stream, stand out against the muted background. These grade control structures, designed to control stream flow and prevent "lateral and vertical adjustment," are the clear calling card of a Rosgen-style approach to stream restoration. "What a waste of money!" Merritts blurts out. "Think of what they could have done with that?!"

The discussion circles back to red polygons on the screen whose impact they are now attempting to summarize. The process of summation involves both locating the historic dam sites and processing the most recent satellite data. The former indicates where this legacy sediment is likely to be located. The latter can indicate the rate of bank retreat and, consequently, the amount of sediment that has been lost in the years between the most recent light detection and ranging (LIDAR) flights. The numbers Lewis has come up with are nearly quadruple what was thought to be coming from these channels. This information may be alarming and unwelcome news for the state of Chesapeake Bay restoration. Within the group, this information brings concern but also a satisfying affirmation of their theories. The challenge now is to convince others.

CONTROVERSY

Despite the apparent success of an earlier project, Merritts and Walter's interpretation of streams is still not fully welcome. Calling attention to a period predating the stream-restoration experts presented a challenge. Clearly stating that this was the scientifically proper restoration target raised hackles. The fact that it ignored the current orthodoxy of stream geomorphology embodied by the teachings of Dave Rosgen struck some as impertinent. *Valley dredging* was a derogative term thrown their way by one well-known Rosgen-trained practitioner who saw their approach as too extreme, unskilled, and impractical. Rather than careful sculpting of channels with appropriate geometry, Merritts and Walter wanted this sediment gone. All of it.

Laced into the criticism was a type of incredulity. How could an academic couple from a small liberal arts college have the authority to call for the removal of thousands of tons of sediment along hundreds of miles of streams? Where would all this sediment go? How could anyone afford this? For homes with stream frontage, how would this affect property values? Maybe this could work in a farm field, but for suburban or urban areas it would simply prove to be impractical. In short, who the hell did they think they were?

Adding to this controversy was the environmental orthodoxy surrounding tree planting. Tree planting along streams has long been a fully accepted and embraced restoration activity. Tree planting can be done at extremely low cost, and volunteers are easily engaged in the project. Riparian tree buffers also have many benefits. They stabilize stream banks and provide shade for streams. This shading lowers stream temperatures, which helps support desired fish communities. Last but not least, these buffers can capture overland flows of sediment and nutrients from neighboring farms or developed lands. In the environmental community, trees next to streams just seem right. It is one instance when careful analysis can be left at the office, and people can agree to just get to it and plant some trees.

Merritts and Walter threw some cold water on this beloved orthodoxy. They described a precolonial stream that was flanked by a mix of shrubs, sedges, and a handful of trees rather than a fifty-foot-wide swath of hardwoods. Not only were these forest buffers historically incorrect, but efforts to create them had often not been effective. They did not hesitate to note that in one ambitious tree planting project of 3,800 saplings, only 12 survived. They explained that legacy sediment set the conditions for incised streams. Incised streams lowered the water table and left the roots of the trees parched. After restoration, the unearthed groundwater seeps would provide greater cooling than any forest buffer could (figure 3.3). The purported value of shade from trees was at best overestimated, and at worst, just wrong. If these challenges were not pointed enough, in many cases the approach they recommended required the wholesale removal of established riparian tree buffers. To achieve the correct restoration target, they said you have to sacrifice some trees.

Another reason that many involved in watershed restoration have been slow to warm to Walter and Merritts's work stems from questions about the implications of this work. Throughout the country, the impact of uncontrolled stormwater on the water quality of our streams has been vast. Roadways, parking lots, and buildings collectively create the immovable juggernaut of an impervious surface. A different sort of legacy, this legacy of development in all its forms, has fundamentally changed how water flows through our streams.

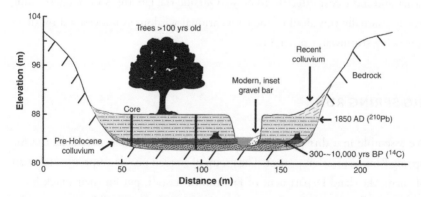

3.3 A conceptual model based on composite stratigraphy from multiple sites, including stream bank exposures, trenches, and cores.
Courtesy of Merritts and Walter.

Rainwater that once was captured by trees or soils with deep-rooted vegetation now runs rapidly across these surfaces into our efficient stormwater pipe network. Occasionally it is routed through filtering devices or retained in stormwater ponds. But in areas developed more than thirty years ago, this stormwater runs directly into our streams untreated. The new "flashy" hydrology wreaks havoc on our waterways.

Many people have worked diligently to reverse these major hydrological changes. People have been taught to "keep rain where it falls." Cities and counties are currently investing millions of dollars on projects to reduce this impact to some degree. This message has been pushed steadily by so many for the past ten years that it has finally, perhaps, reached a state of general acceptance.

Walter and Merritts's work doesn't ignore the reality of altered hydrology, but they are emphatic that the primary reason for the widespread channel erosion of Lancaster County streams is not flashy stormwater flows that plague many urban and suburban streams. They point their finger at the existence of unconsolidated legacy sediments. In a world where legions of well-intentioned professionals are finally seeing the stormwater message sink in, Walter and Merritts are saying, "not so fast."

Implicit in this attribution is a suggestion that resources might be better steered toward projects that remove legacy sediment and create functional riparian wetlands that can, over the long-term, capture a range of pollutants that have made it into the stream. The pool of restoration dollars is limited, and if effective treatment can be realized through one stream-restoration project, it is likely easier and more cost-effective to do so. Outside the lab, this sort of questioning has occasionally run afoul of the conventional wisdom as understood by those advancing stormwater programs.

BIG SPRING RUN

We assemble in a dirt parking lot next to a large barn that overlooks the Big Spring Run restoration project. The group of sixteen people, approximately half of them Maryland Department of Environment staff, put on their muck boots and sip coffee as we wait for the tour to begin. Walter gets things started by handing out a fact sheet and explaining the schedule. He encourages everyone to walk around the site and explore because the time for the tour is relatively short. We'll reconvene in their lab after the tour and can dive into any questions further at that time.

We pass through the orderly white picket fences that surround the farm and cross the road to the stream. From this elevated vantage point, the group has unimpeded views of the entire restoration site. We look out at a one-hundred-yard-wide band of tall grasses and dense vegetation surrounded by a bucolic landscape of gently rolling fields. On the horizon, swing sets sit in expansive yards that have been carved out of former farmland.

It is late in October, and the vegetation has lost its green lushness but remains dense with plant biomass. Red-winged blackbirds screech their familiar call. Stream channels are not visible except for short sections that pop out from the dense mass of cattails and sedges. As Walter explains the project, Merritts slips away into the cattails. She is carrying a surveying rod and using it to plow through the wetland.

In 2011, 22,000 tons of sediment were removed from a 1,500-foot section of this valley. At least 1,500 trips by double-axle dump trucks were necessary to rectify the impacts of more than one-hundred years of erosion.[2] This removal allowed for the creation of 4.7 acres of riparian wetlands. The restored stream measures more than 6,000 linear feet—double the original length, accounting for sinuosity and multiple channels. In returning to a restoration target earlier than any of our personal histories, ecological function was expanded beyond what would be considered possible.

In accounting for restoration work, rigid categories have been formed to help tracking and assessment. Stream-restoration projects can be compared with other stream-restoration projects; wetland-restoration projects with other wetland-restoration projects. Walter describes this project as neither a stream nor wetlands restoration project, but both: "It's a square peg in a round hole." This unique status makes it difficult to compare these projects with other projects in terms of sediment and nutrient reduction or on a cost-effectiveness basis. Accounting is important because it sets cost-efficiencies that increasingly drive funding decisions. Most stream projects earn "credits" based on the newly created stream's ability to process nutrients and capture sediment. Previously set at one rate per linear foot of restored stream, recently developed reduction rates try to account for several different restoration approaches.[3] Walter and Merritts are trying to make the case that those 22,000 tons should also be counted because, barring restoration, they would have been washed down to the bay eventually. Not everyone is willing to give them those 22,000 tons.

Ten minutes later Merritts reemerges thirty yards downstream. She holds the surveying rod at shoulder height as she stands next to a black-and-white survey pole. "This was the height of the legacy sediment prior to the restoration project," she shouts from the cattails. The space opened by the removal of this thick layer of soil is a visual testament to the relevance of their research.

As people begin to process this massive amount of earthmoving, a key backer of the project begins talking. Jeff Hartranft, an ecologist and manager for the Pennsylvania Department of Environmental Protection, helped obtain funding for the project and shepherded it through uncertain permitting. He is sporting a Philadelphia Eagles shirt and hat, camo muck boots, and the ubiquitous government ID lanyard. As he holds a map in his hand, he dives into the bottom-line currency of these projects—pollutant reductions. This project has been heavily monitored to better understand the reductions they provide. Previously a site for a U.S. Geological Survey gauging station, the monitoring benefited from ten years of prerestoration data. Depending on the way flows are calculated, the project has been shown to reduce between 70 and 90 percent of suspended sediment in the stream.[4] This doesn't account for the 22,000 tons of legacy sediment that has been removed.

Water temperatures have changed for the better as well. Following completion of the project, small springs have been observed in various places throughout the valley. The subterranean hyporheic zone that conveys water between the stream and localized groundwater has been reactivated. Hartranft tells us that this groundwater seeping into the stream has lowered peak summer temperatures by 20 degrees Fahrenheit. He confidently states that we don't need trees to shade the stream because the shrubs and taller grasses are doing that just fine.

Moving on from the water quality benefits, Hartranft dives into his passion: plants. He talks about the state-threatened *Juncus torrii* that is now present. He mentions that thirty species of plants in the mustard family exist here now, up from the eight that were formerly found here. He says that the sedge hummocks that are forming throughout the stream/wetland complex are providing a highly valued and stable microtopography throughout the site. At the outset, most people said the precolonial vegetation restoration target was impossible, and he seems pleased to relay that we got "pretty darn close." Reflecting on how restoration is mostly science-based but also possesses some alchemy, he tells us that the threatened *Juncus torrii* wasn't even in the seed mix and "must have come in with the waterfowl." But perhaps most revealing is that this wetland has reached the "exceptional value" class in the technocratic classification of wetlands. This class makes up only 5 percent of natural wetlands in the state. As the mic drop moment of his presentation, he declares that he's never before in his career seen a restored wetland hit this high of a mark.

For most people in this field, achieving the results that are evident here would be satisfying enough. One could certainly call it a day and produce some impressive fact sheets. As the assembled group breaks off into individual discussions, one person is still out in the wetland. Merritts is deep into the interior of Big Spring Run, taking photographs of side channels to document how the stream may be evolving. Walter informs us that we need to leave as he glances back at the wetland. Rhanis heads over toward the edge of the newly liberated soil and calls Merritts back in, evidently familiar with this task.

LAB PRESENTATION: GEOSPATIAL ANALYSIS TEEING UP RESTORATION PROJECTS

We're back in Walter and Merritts's lab waiting for the presentation to begin. Lewis's PowerPoint presentation crashed earlier this morning, so Merritts provided reassurance and a substitute presentation that covered much of the same material. People are picking through an assortment of Dunkin' donuts, pouring coffee, or checking their email. Walter gets people focused on the screen and dims the lights. Lewis begins by showing the maps of erosional areas and explaining their process for identifying them. Most people are familiar with the history of the mill dams, so he focuses on the highlights. More than 691 former dam locations have been identified in Lancaster County alone through a combination of poring over historical maps (figure 3.4), a detailed examination of aerial imagery, and field reconnaissance.[5]

He explains how Rhanis developed a script that can determine both depositional areas and the erosional zones. He describes how using remote sensing sidesteps an insane amount of fieldwork that would be required to gain similar information. In many cases, fieldwork is not more accurate. One common method of measuring bank erosion that involves sticking smooth pins into banks and regularly measuring the length of the pin is particularly problematic. During a recent return visit to one of their sites last spring, they found their three-foot

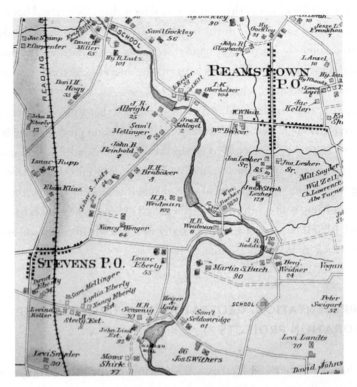

3.4 Cocalico Creek in Lancaster County in 1899. Note that mill ponds had reshaped the creek at regular intervals by this time.

Atlas of Surveys of the County of Lancaster, State of Pennsylvania, Graves and Steinbarger, 1899.

bank pins splayed among a collapsed bank, the volume of which could fill a few minivans. This seasonal calving of stream banks, called the freeze-thaw process, is a phenomenon that was first described by Reds Wolman. So much for hands-on measurement.

The more pedestrian, ground-level process for identifying stream-restoration sites has been practiced for decades. It relies on the local and professional knowledge of any number of people who have some direct experience with the local streams. It is a time-consuming physical assessment carried out by individuals, including conservation district staff, researchers, and landowners. Yet as soon as the first large storm rages, this channel assessment is out of date. Projects are selected based on incomplete or out-of-date information. This process has been described by those familiar with its limitations as "driving around and hoping for the best," or somewhat cynically, "random acts of conservation."

Rhanis takes his turn at the front of the room. He begins explaining his back-end remote sensing work. He describes the various LIDAR flights with the same scrutiny as a wine snob assessing the latest vintage of a pinot noir. "The 2014 flight meets the QL2 standard; however, the 2008 doesn't truly support that cell size," he states with an apologetic tone. Rhanis speaks carefully and clearly, making sure not to make any false statements. If his words were transcribed directly into a science article, it would pass editorial review. He mentions "zonal statistics" and "threshold levels." Apparently, road crossings over streams create artificial dams that create false flow paths that mess up the analysis. This painstaking task of digitizing flow paths is Rhanis's current focus. If he's successful, his script will correct for these errors and automatically calculate erosion amounts for any two aerial flights, which could be applied anywhere there are quality LIDAR flights. Rhanis continues with some mention of the Digital Elevation Model (DEM) differencing and avoidance of extrapolation of high error. Although it's not certain that everyone else understands, it's pretty clear that Rhanis does. Wrapping up, he shows the map of the red polygons again and tries to put it all in context. He points out several large red polygons along one particular stream and, loosening up, chuckles, "these are ones that knock you over your head saying 'please restore me.'"

I survey the crowd, and several heads are either down on a table or nodding toward it. The work of the Franklin and Marshall College team has the potential to identify the exact locations of the most highly eroding banks and provide an estimate of the amount of sediment loss that can be averted. In the hands of funders and regulators, this information could create the foundation for a scientifically based optimization for all future stream-restoration projects. In an era in which financial resources are limited and clear progress is demanded by the public, this research could be transformational. However, whether it's due to the level of detail or falling caffeine levels, it appears that not all are ready to grasp the importance of this research.

BRUBAKER RUN: WIN-WIN-WINS FOR STREAMS, STORMWATER, AND DEVELOPMENT

Our final stop of the day is a highly consequential legacy sediment stream and wetland restoration project in the middle of an unremarkable area. The surrounding area includes health care outpatient clinics, low-rise apartment buildings, and nondescript commercial development. It's the kind of sprawling development that goes unnoticed due to its ubiquity. The stream-restoration project that exists at its center is notable because it was pursued in large part to help the developer meet his sizable stormwater requirements. The project was conceived of by Justin Spangler of Landstudies, a consultant who has worked closely with Walter and Merritts on constructing Big Spring Run and other legacy sediment removal projects in the area.

Spangler stands with his back to the expansive restored wetland and stream complex. The wetland bordering the stream creates a 200-foot-wide corridor of functioning habitat below the larger Lime Spring Square development. Spangler self-consciously mentions the "apologetically sinuous stream," aware that the single-strand sine curve stream channel has become a sign of hubris. The idea of what a channel is and should be is currently changing. Spangler explains that they are anticipating that beneficial side channels will form over time.

Spangler breaks down the stats: 4,700 feet in length, 8.4 additional acres of wetland created, 38,500 tons of legacy sediment removed. This project will prevent 400 tons of sediment and 254 pounds of phosphorus from reaching Chesapeake Bay each year. The rote recitation of these numbers is standard practice and provides an almost liturgical comfort. These numbers are impressive, but 85 percent is the number required for this project to get approval and make many public works managers convert. That is, this project alone satisfied 85 percent of the stormwater permit requirements for the entire East Hempfield Township with no cost to the taxpayers. In some cases, the requirements of these five-year permits can require hundreds of thousands of dollars in stormwater retrofit projects in addition to significant staff time to implement better street sweeping regimens or better pollution prevention programs. The subtext is clear, if unstated. The project has provided ecological uplift, met regulatory obligations, and saved money. This is the win-win-win that all projects are shooting for and what may truly drive the pace of future legacy sediment removal efforts.

This project delivers this victory because the 8.4 acres of wetland created along the channel are serving a function. Beyond providing habitat for wildlife, these wetlands are a secondary treatment for the vast amounts of stormwater that runs off this sea of pavement that allows us to drive to every service we require. It also helps process the higher flows from upstream when they reach the project site, spread out to cover the floodplain wetland, and drop their pollutants. This design uses natural features to address human caused pollution. In the process it sidesteps a traditional player in this field.

The stormwater management industry is a multi-billion-dollar industry that has been built to accommodate developments like this one. A host of best management practices (BMPs) have been designed and patented to be plopped into the corners of these developments. Stormwater ponds, catch basin inserts, baffles, and oil and grit separators are a few of the structures used. The combination of practices can be tweaked to meet specific concerns about specific pollutants. A legion of stormwater engineers can guide you down this selection path. An entire subsector of the field focuses on proper maintenance of these devices.

In the Brubaker Run project, the extent of traditional stormwater management practices has been shrunk down to one small stormwater pond next to the parking lot. This small, depressed area serves as the first treatment component of the overall project. Motioning to the wetland below, Spangler exclaims, "We're not dumping dirty water down there!" It's a question no one asked but one he wants to address. The concept of using natural ecological features such as streams and wetlands to treat polluted stormwater runs afoul of the central paradigm of most water regulations—treat your pollution prior to discharging into nature. Regardless of the number of hyphenated wins, this approach raises the hackles of some. Every biologist and most regulators cringe at the thought of using natural features to treat our pollution. Also, it is unlikely that the traditional stormwater industry will warm to this apparent short-circuiting of their hegemony.

What remains to be seen is how the public reacts to this project. People despise mosquitoes. Wetlands equal mosquitoes. Answering another question that wasn't raised, Spangler mentions how the birds are controlling the mosquito population: "I haven't been bitten yet!" Spangler keeps piling on the wins. A retirement home located adjacent to a similar project nearby charged more for a room with a view of the stream. Last but not least, a portion of the 38,000 cubic yards found its way to a brownfield project in Lancaster, the soil serving as black gold for landscapers. With such an array of successes, Spangler envisions more benefits farther afield. Acknowledging the high cost of truck transport of this massive volume, he suggests, "Why not put the legacy sediment on train cars and ship it up to West Virginia to the strip mine sites?" The notion of current restoration projects feeding other mine reclamation sites sounds fantastic. The historic propriety of soil returning to an upland source not far removed from its original location 250 years ago is so appealing I can't shake the thought. The fact that it all makes so much sense makes me wonder what the catch is.

After Spangler finishes his presentation, the group dissembles and wanders out into the 8.5-acre site in pairs and small groups. I catch up with Merritts who had made her way out fifteen minutes earlier. She is back in the wetland, a human-sized king rail poking around in the side channels. When I reach her, she points to a line of accumulated drift that has been placed there by recent high water. It was good to see this drift so far away from the channel with no signs of scouring. The stream was functioning as it had 300 years ago.

Despite the fact that she'd probably seen this site two dozen times and given nearly as many presentations on the project, her energy for scientific inquiry never seems to wane. I wonder what she thinks of the integration of her and Walter's work into the regulatory compliance world. This project shows how returning a stream channel to precolonial form can work extremely well within our current regulatory framework. But this convergence of restoration and regulation raises new questions. The fact that this ecologically and geologically correct stream project was created in part to facilitate twenty-first-century suburban sprawl presents an uncomfortable irony.

As I consider how regulatory and economic forces might work together to return streams to their proper form and function, I think back to Merritts's impassioned description of boulders transported by geofluction rather than by stream forces. This ability to understand the geological processes that have played out over eons provides an expansive perspective from which to assess our current trajectory.

Personally, I seek that feeling of hope I felt on my first trip to Lancaster County. The once buried organic soil horizon, liberated after 200 years, now supports a diversity of flora and fauna that should exist here. A river provides the expansive ecological benefits that we had forgotten they could provide, a sense of ecological and historic correctness.

The people involved in the legacy sediment work in Lancaster County have checked all of the boxes. Now that I know more, I find that I have more questions. I wonder when and how the economics of the project will work out. I wonder if we are we missing any root causes with this disciplined focus on nutrient and sediment reductions. Could this approach save us from our impervious expansive footprint, or perversely facilitate more of the same? Standing here in this developing stream wetland complex in this new development, I have some discomfort with the idea that this restored stream is serving humans so directly. For all I have learned from this academic couple, I want to know more. I imagine that is what any good teacher strives to accomplish.

CHAPTER 4

THE HUMAN BEAVER

Mega- to Micro-Engineering Solutions in Greater Cincinnati

I had been planning my trip to the Queen City for more than a year, juggling the challenges a global pandemic brings to a travel dependent investigative project. The initial seed had been planted a couple of years earlier. Two stream projects, radically different from each other, caught my attention when I attended a conference in Cincinnati. The first was one of the most ambitious stream daylighting projects ever undertaken. It promised to return a stream to a part of the city that boomed with early manufacturing and industry and then, over the past thirty years, faded into relative poverty and blight. The project promised to help the city meet its EPA mandated requirement of reducing combined sewer overflows at a lower cost than traditional methods. The second project was low-cost and low-tech. It involved physically jamming and jiggering logs and larger branches into small headwater streams besieged by uncontrolled stormwater. The goal of this unvarnished approach was to reset a stream and, given enough time, hope that the stream might repair itself. I soon learned that the same person was behind both projects.

I had spoken to Bob Hawley on the phone a couple of times and discovered that he was the stream designer for the much-discussed Lick Run daylighting project. Fortunately, he was planning a log-jam installation just outside of Cincinnati in Milford, Ohio, in July. He invited me to come to this installation on a small, unnamed creek that originates just outside of the beltway, flows down through the Cincinnati Nature Center, and eventually flows into the east fork of the Little Miami River. He intended to complete the project in just two days and said I was welcome to take part in the entire process. The partners in the project included the Clermont County Soil and Water Conservation District as well as the county's Water Resource Department. Some Parks Department maintenance staff were going to help as well. The head of the conservation district was interested in the method, and Hawley would be training the staff on the techniques. Although it seemed like he had plenty of labor, I packed my boots and work gloves along with my notebook. If I was going to be embedded in Hawley's volunteer army, I wanted to be able to pitch in if needed.

In a field dominated by people pitching their silver bullet, I was curious how someone was able to apply widely divergent approaches in dramatically different situations. This challenged my preconceptions of engineers pulling out spec sheets from the appropriate manuals and rigidly applying them toward the identified problem, even though the challenges faced nearly always emerged from a set of watershed conditions that required an expansive intervention of interconnected solutions. I wondered whether two projects at opposite sides of the budget spectrum would shed light on the questions regarding cost-efficiency. Are millions necessary for the realization of meaningful projects? Can anything meaningful be accomplished with small budgets?

MANNING UP IN AN UNNAMED CREEK:
WHAT 10K AND A LITTLE MUSCLE CAN ACCOMPLISH

I had driven down the day before. The uneventful drive across the fertile plains of Indiana became interesting when I descended into the Ohio River valley. The lush green hills radiated heat and humidity and reminded me of a rainforest when compared to breezy Wisconsin. The feeling was solidified after a series of monsoon rainstorms blew through and forced me to pull over at several points. Nerves frayed, I was happy to arrive in one piece.

The following morning, I drive to our meeting spot. As I pull off the beltway and head toward Shor Park, the entire landscape transpires water vapor. I pass through ubiquitous exurban developments. Ranch and two-story homes built in the 1950s crowd out the occasional original structures of the white colonizers of the Ohio River valley.

This area east of Cincinnati was settled in the 1780s, before the settlement of Cincinnati proper. George Washington himself surveyed this area, and following the Revolutionary War veterans were given plots along the Little Miami River for their valor and service. They traveled downstream on the Ohio River, from Pittsburgh and other upriver towns, to claim these plots demarcated by the man who would become our president.

Nearly 250 years later, I wonder whether Washington might now regret handing out those free parcels of land. He would recognize the major water bodies
but would be hard pressed to understand the present-day land use. New development is comprised of strip mall–type businesses that house conveniences for
commuters who live in this mostly rural area. Along the winding roads, a vista
occasionally presents a 150-year-old house perched on a rolling hill and framed
by a bucolic field. It reminds me that this area has been settled for a long time,
and for those who possessed resources, these river towns could offer a grand life
indeed. In 2021, the landscape is mostly unremarkable. It resembles any number
of places outside larger cities where land is cheaper, taxes generally lower, and
space is abundant.

I pull into the parking lot of the county park—a tidy park that includes a
newer pavilion with a jaunty blue metal roof, a modern version of a cupola, and
beige stonework reminiscent of wealthy subdivisions. New play structures and
fields of turf surround the building. In the roundabout at the entrance, an American flag and a small windmill lie motionless in the still air.

When I meet Hawley in the parking lot, he greets me with a smile and welcomes me. He's sporting a pink hat that advertises the more pristine locale of
Lake Tahoe. He's bearded, slender, and about five-ten, yet he exudes the sinewy
strength of a wrestler. He and his project manager, Peter Tower, are unloading
supplies from the bed of their truck. They drove in this morning from Lexington,
Kentucky, where they both live. At this site, Hawley is meeting with the client, a
handful of people from local organizations who are curious about the project, a
logging crew, and me. Hawley doesn't seem phased by having to juggle this range
of interests and personalities. He telegraphs an easy preparedness that instills
confidence.

After he and Tower grab what they need, we head down a path into the woods
toward the small, unnamed stream. The stream winds through a mixed stand of
trees and shrubs of the variety that quickly establish in formerly cleared areas.
Twenty-foot-high sycamores, invasive autumn olive shrubs, and box elders have
colonized the floodplain and blanket it with dense shade. Honeysuckle shrubs
anchor the banks in several locations, and poison ivy lurks in the tall grass. The
stream itself lies about three feet below the floodplain and possesses a grayish
tint from the residual stormwater flows the previous day. The eroded banks and
exposed tree roots tell me that this stream can quickly transform into a torrent.
We waste no time and head straight toward the stream.

Hawley leads Tower and me to a particular bend where he intends to install his first structure. There's a sizable sycamore on one bank, and on the opposite side is the trunk of a tulip poplar. Hawley tells me this is a perfect spot to lodge his key logs. The subterranean root network of these two trees appears to anchor a substantial section of the stream bank on each side, but given the active erosion above and below this spot, these trees seem to have a precarious existence. Their forty years of tenure with the land below could end after another large storm. Hawley indicates places along the banks where logs can be jammed to ensure they can withstand these flows (figure 4.1). Up against this root, in between those two branches, Hawley preps us. Staring at this stream it seems logical, but the logistics aren't yet clear to me.

Things start to come into focus when Hawley points up the hill on the opposite side of the bank where three older guys are cutting down young sycamore and locust trees and sliding them down a steep bank. The area has become a makeshift logging site with the hillside conveniently acting as the skid pad.

4.1 Bob Hawley installing a log structure in a tributary of the East Fork of the Little Miami River in Milford, Ohio.

The setup is ideal; the logs slide down the steep slope and land about twenty feet from the stream. Hawley seems happy with the arrangement and mentions that this is the best setup thus far for one of these projects. Clearly there has been some advanced planning.

We walk down the valley and find the staff from the two county agencies. Hawley introduces me to the group comprised of three women and a middle-aged man who oversees one of the departments. This is definitely not a standard workday, but they are cautiously excited about being here, acting as clients, day-laborers, and students. We share some small talk and await Hawley's direction. After he confers with the loggers, Hawley leads us back to the bend in the stream we had just scouted.

He tells Tower that the three-to-four-inch diameter sycamores are perfect for this first structure. We follow Tower to a handful of small trees that have been felled and delimbed expressly for our work. These smaller logs are easy for two to carry, so we pair up and start hauling them to the stream. We walk somewhat gingerly over a floodplain that is pocketed with tripping hazards. Once we get four logs to the spot where Hawley wants to start, the two of them pick up a fifteen-foot log and start to angle it into a small spot between two roots of the tulip poplar. They secure this end and then rotate the other into the exposed roots of the sycamore on the opposite bank. Hawley directs the effort with familiarity even though particulars of this stream and available materials are unique. Once installed, the logic behind the placement reveals itself. Lodged between two or three fixed points, the log is braced to handle any future flows. Hawley and Tower continue with another key log, but this time they angle it in roughly the opposite direction, positioning the high end of the log on the other bank so the two form a sort of "X" in the channel. They then lay a couple of heavy logs on top of those two, using the weight of both to secure the first two logs. They continue installing logs, threading them through the logs that have already been installed, creating new angles and planes that push the structure into an indescribable non-Euclidean geometry.

Hawley and Tower are working hard, and our makeshift crew appears antsy to help. I make an obvious beaver-themed joke. Despite not exactly landing the punchline, I get a couple of polite smiles. Given the complexity, the work requires focus. The work is steady and full of unspoken purpose. The sort of work at which rodents excel: tug, pull, shove, chew; repeat.

While Tower and Hawley are jamming logs into this increasingly impressive structure, another person has joined our informal work crew. Adam Leyman, the director of the Soil and Water Conservation District for neighboring Hamilton County, is taking it all in. He looks at the logs angled across the channel and tentatively asks Hawley, "So this is the . . . design?" "This is one iteration of the design," Hawley replies, his measured response indicating that he has been asked this question before. Leyman nods, uncertain what to make of Hawley's answer.

Despite the frankness with which he replies, Hawley's loosey-goosey explanation is nothing less than radical for this field. When stream designers explain their approach, it almost always involves numerous technical terms, a declaration of a particular standard or approach, and references to modeling that involves math we wouldn't understand. A pageant of credentialism is expected. Although he means no ill will, Leyman's question signals a trained skepticism that any public servant believes is required. It also indicates that he's been around a bit and met other designers. It's a raised eyebrow and an unspoken challenge that permeates nearly all discussions about stream restoration. It says, "impress me."

The problem with this dynamic is that it creates immense amounts of pretense. It can add kerosene to a bonfire of toxic masculinity that can get out of hand when so many men are competing over scarce resources. But most important, it frequently prevents honest discussion about what is needed and what is important for any specific project. The restoration needs for any particular stream section vary as widely as the shape and form of the streams themselves. The only similarity between a headwater stream in suburban Cincinnati and a Louisiana slough is that water is found in both. The process of standardization that undergirds a typical engineering design process often leads to an expectation that a stream-restoration designer will have a universal, transferable, and optimal solution for every situation.

Hawley seems to have little interest in this game. He and Tower are positioning these logs in a fashion that works in this particular place. It makes no sense to spend hours surveying the entire stream; determining what sorts of logs are ideal versus what types are available; calculating the ideal length to cut said logs; and then fiddle with a representation of the actual stream on AutoCAD to determine where each log should reside. This exercise is followed by submitting these plans to others for technical review, comments, and changes. The labor cost implicit in this process, not to mention the reams of paper required, drive up the cost per linear foot for any restoration project before any dirt has been moved.

But this is the process most stream-restoration projects follow. Hawley mentions that they've been able to avoid this lengthy process, as well as the cumbersome and costly permits associated with them, for these types of projects. "So far," he adds. Hawley and Tower continue their iterative placement of logs, in the process highlighting the value of field knowledge.

Returning our attention to the newly created structure, Hawley explains, "next we just need to weave these smaller branches into these key logs." Pointing to a bank that is actively eroding, he tells us, "Use the small branches on the areas closer to the bank. This will become a depositional zone and fill in." After he weaves in a longer branch with leaves still attached, we start to see what Hawley's seeing. The bank that was regularly hammered by high flows is now protected by this mass of logs and branches. The jumble of biomass seems likely to catch any branches that may flow down, and eventually it will catch sediment running off the construction sites I drove past on my way here. This is the idea, Hawley explains to us; setting up the conditions for the stream to once again start trapping wood and leaves and, in due time, the sediment and attached phosphorus that courses down toward the east fork of the Little Miami River.

After this introduction, Hawley encourages Leyman and the others to begin weaving branches into the key log structure. It's a job that requires the energy of a ten-year-old and the stamina of a mule team. Leyman appears up for the challenge. I tag along with Hawley and Tower as they set off downstream to the next installation site that lies around two sharp bends.

As we head over, I duck under a tree held at a 45 degree angle thanks to a sizable grape vine hanging from its canopy. Even the forest seems to be engaging in this process of woody entanglement. Once there, we assess the stream channel. It's another spot where a large tree secures one bank and a tough honeysuckle shrub with multiple stems anchors the opposite bank. The site is convenient because it lies not far from the base of a slope that has been used to transport logs. Hawley and I survey the pile of logs and Tower picks through them. The relatively light, four-inch diameter sycamore logs are gone; only saturated seven-inch diameter locust logs remain. Hawley tells me the locust logs had been cut a couple of days ago; and yesterday's gusher caused them to become waterlogged and heavy. Tower dislodges the smallest one that still has adequate length.

At first glance, the log seems pretty manageable, especially for three of us. But as soon as we try to pry it from the soft ground, it's clear that it's not going to be that simple. The log must weigh at least five hundred pounds, and walking over the uneven terrain means that each person will get unexpected loads. As Tower navigates up a slight rise in the floodplain, I find that the log magically lifts off my shoulder. A minute later, when he shimmies down the slippery bank, nearly all of the weight instantly bears down on my shoulder. I try to step on rocks in the stream to avoid completely flooding my boots while also trying to keep in line. The log seems to be dictating the path. We keep pushing ahead like drunken mercenaries intent on battering down the castle doors.

As the log grinds into my shoulder, I look for a way out of this test of manhood or whatever this is. I use my status as an embedded reporter to ask a question that will, conveniently, force us to take a break. The question is one that few men, including me, are wont to discuss. I ask Hawley why he thinks the field is so dominated by men. Hawley is unflappable. As he wipes his face, he leaves a streak of mud under his eye. "There are actually a lot of women in the field doing important work. Ellen Wohl out west, Margaret Palmer and Dorothy Merritts in the east. And Rebecca Lave is putting out a lot of papers." He mentions that he recently hired a woman PhD scientist who came from academia and has become involved in implementation. "It's great to have someone well-versed in the science side of things. I really appreciate having someone who can help me keep tabs on the latest research—it's a bit challenging while trying to run a business." Although Hawley reframed the question, his emotionally intelligent answer makes me feel good. There are many women doing important work in this field. Yet this enlightened feeling is tempered by the knowledge that we haven't finished hauling five-hundred-pound logs around in a pock-holed floodplain, broken ankles be damned.

I probe a bit about the business side of things. I know that it's a tough business in which to succeed. Hawley says that the stream projects make up about half of his business and are growing. Stormwater designs and retrofits make up the other half. He even has a patent on one particular stormwater pond retrofit that allows only two-year storms to pass through. This type of mechanical installation was common, but Hawley added a component that minimized the damage from larger storms. His comfort with a range of projects fits well in a field that is increasingly trying different approaches. It also fits with my image of what an engineer should be—able to figure out a solution for any situation.

I try to understand what makes certain people go out on their own in this highly competitive field. Hawley mentions that the Lick Run project where he obtained a subcontract from the prime contractor gave him the start he needed.

After he got that, "I knew that I could eat," he says. He also was able to get health insurance through his wife's job as a physician. Since then, he's built the business up to seven staff and one paid intern. There were key points along the way.

Hawley mentions one key juncture when he was presenting at the 2018 Ecostream: Stream Ecology and Restoration Conference in North Carolina. He had recently published a paper in *BioScience* that was critical of how many stream-restoration projects tried to re-create a specific channel form without adequate analysis of the valley soil materials and discharge. His article states that a "plurality of United States–based stream channel designers (perhaps even a majority?) organize their designs around three well-intended but fallible practices: regional curve dimensions, Rosgen planform pattern, and grade control structures to constrain the profile (i.e., dimension, pattern, and profile)."[1] Those three terms—dimension, pattern, and profile—are the "Father, Son, and Holy Ghost" of any Rosgen class and have been drilled into the heads of tens of thousands of students. To call them out was a direct challenge to the orthodoxy of stream restoration at the time, as well as an implied challenge to the man who trained so many of these designers.

When he arrived at the conference, a friend mentioned that Rosgen was there and had a "burr in his saddle about Bob's paper." This was one of Hawley's first presentations as an independent stream-restoration designer. Rosgen had already designed hundreds of stream-restoration projects and had trained more than 10,000 people in the field. Hawley mentions that Rosgen did indeed attend his talk and sat in the front row, cowboy hat on and arms crossed for the entire talk. Tumbleweed bounced down the hallways of the conference center.

After the talk Rosgen asked a few questions but then stuck around for some extended follow-up. The discussion was far from a casual exchange of ideas because Hawley's most important client happened to be right there, passively observing the tête-à-tête. Reputations are made and lost during interactions like these. One must be confident but not overconfident. One must show independence but also awareness of prior knowledge. This is even more relevant when you're talking to someone dressed up like Gary Cooper in *High Noon*. Hawley said they had talked for more than twenty minutes and that it was not a particularly easy conversation. "Eventually we agreed on all points," he laughs, the pressure of the moment now far removed. He said Rosgen was pointed but not rude. Rosgen claimed rightly that, in his classes, he does encourage the sort of valley analysis that Hawley highlighted in his paper. But Rosgen agrees that some people trained by him had cut some corners. This train of thought is common among many in the restoration field. Rosgen knows what he's talking about, but some of his trainees have dropped the ball. It's a comfortable place to land, especially when you're addressing the cowpoke that created the entire field and trained almost everyone in it. Some people get it wrong, regrettably but understandably, because, well, they're not Rosgen.

When we've finished the second installation, we return to the other team who are presently weaving smaller branches through the log framework that Hawley and Tower installed. Leyman, the conservation district director who had asked about the log-jam design, is enthusiastically wrapping smaller branches through the exposed roots of a streamside tree. Although he is silent, he is enjoying the process. He moves in small quick movements, tugging and pushing. He is focused and intent. The excited yelp of a dog, or perhaps a coyote, breaks the quietude, and Leyman abruptly stops his work. He turns to look at me, freezes for a moment, then returns to work.

ALL IN FOR MILL CREEK: A QUARTER BILLION FOR COMBINED SEWER OVERFLOW (CSO) RELIEF, STREAM DAYLIGHTING, AND COMMUNITY AMENITIES

The next day I decide to swing by Hawley's marquee project before revisiting our worksite at the unnamed creek. I am returning to a site I visited three years ago while attending a conference in Cincinnati. At that time, it was a major construction site occupied by backhoes, dump trucks, and a lot of dust. Like a patient on the operating table, the twenty-foot-diameter brick combined sewer pipe that had contained the stream for more than a hundred years had been exposed, revealing the ubiquitous but usually invisible infrastructure that undergirds our cities.

Excitement surrounded the project. It was a daylighting project, a project that unearths a stream and brings it back to life. The resonant metaphor implicit in these projects makes them irresistible to people of every stripe. Ambitious and attractive, they are the supermodel of the restoration field. They also are generally the most expensive and technically challenging. The projected budget, which included several park amenities, came in at $244 million.[2]

The approach is nearly always more complicated than just re-creating a channel in the general area where one had once been. The proposition is not just removing a stream from a pipe but reappropriating a space that had been fundamentally altered by human development. This was true for Lick Run.

From the time it ran as a pristine tributary to Mill Creek, to the period when it became an overflowing and debris-filled ditch, to its subterranean encasement in 1893, to the time of renewal, almost everything about the stream and its watershed had changed. The 2,700-acre watershed that drained into Lick Run had been densely forested when the first German immigrants began to settle in the area. Joseph Niehoff cleared much of the land for his dairy farm when he set up his operation in 1886.[3] A couple of years after piping the stream, he purchased other farmland and moved his cows farther away from the city. With that, Lick Run's brief agricultural period of the late 1800s transitioned to a period of light industry. Throughout the 1900s, federal-style townhomes and duplexes were built to house the people who worked in the nearby factories. Consequently, the imperviousness of the watershed increased. In addition to stormwater runoff, the stream received raw sewage from homes and businesses.

Moving into the modern era, due in part to the more intense storms fueled by climate change and an altered hydrology that sent nearly all rainfall directly to the sewers, combined sewer overflows became more common. This situation of too much stormwater runoff rapidly rushing into the same pipes that carried household sewage meant that this untreated mix would "overflow" at relief points further down in Mill Creek. Solidifying this conversion of land use, Queen City and Westwood Avenues, parallel one-way arterial roads that run through the valley, were constructed, leaving only a 250-foot-wide strip of land between them that was soon filled with houses. The only thing that remained unchanged were the steep contours of the valley. The altered hydrology combined with the historic topography created a watershed that hydrologically acted more like a flashy desert arroyo than a temperate headwater stream.

The primary impetus for undertaking the daylighting project stemmed from these combined sewer overflows. The average amount of untreated sewage released during these events was estimated to be 1.5 billion gallons annually, rendering Mill Creek unsafe for human contact.[4] The number of overflow events each year varied with the rainfall patterns, but one estimate put it at forty each year.[5] Anyone who came in contact with the stream would have greater than a one in ten chance of exposing themselves to untreated sewage. These water quality violations continued for many years before a judge issued a consent decree mandating that the Metropolitan Sewer District of Greater Cincinnati reduce sewer overflows dramatically. The standard solution would be to build extremely large holding tanks that could retain huge volumes of combined run-off, eventually pumping it to the plant for proper treatment. Many cities in the country have adopted this approach, and despite the high initial costs and steep operating costs, water quality improvements have followed. In this case, however, the broader design team and the EPA saw an opportunity to separate the stream and associated stormwater flow from the sewage. It would require building a system for the stream that could handle large rain events without flooding the valley. Given the impervious cover in the watershed, the existing infrastructure at the bottom of the valley, and increasingly severe storms, a relief system would need to be incorporated (figure 4.2).

The design that solved this predicament involved multiple strategies. The hybrid approach would reduce the number of overflows to an annual average of four and reduce the total combined sewer overflow volume by 800 million gallons.[6] First, the city installed several green infrastructure projects that directed stormwater into the ground rather than into the pipe system. These were installed along roadways and parking lots and helped to push back the clock on changes to the watershed hydrology. To ensure that the stream wouldn't flood during large storms, they built a new storm sewer underneath the stream. Overflow grates in the floodplain ensured that high flows in the stream would drain into the storm sewer underneath. They built something unknown in nature, a stream that couldn't flood.

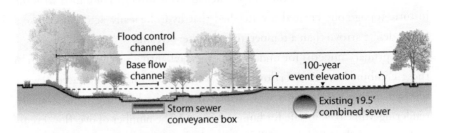

4.2 Cross section showing the new stream channel of Lick Run running on top of the storm sewer conveyance pipe. The stream and the overflow storm sewer divert runoff away from the combined sewer pipe, thereby lessening the frequency of overflows of untreated sewage.

Courtesy of Human Nature, Inc.

The cultural demands placed on such a project had also changed. Before it had been buried, the stream had been a foul ditch that overflowed its banks and left fetid pools for mosquitoes to breed. It was a convenient place for some neighbors to throw their trash. When the community pondered the rebirth of the stream, they hoped for all the services that streams can provide. People wanted to access the stream. Recreational opportunities needed to be incorporated. The Cincinnati Preservation Association highlighted the architectural and cultural significance of many of the federal-style buildings that were slated to be torn down to make way for the restored channel. On the economic side, the community hoped that a portion of the sizable project budget could put local residents to work. And perhaps most important, the stream could not flood under any circumstances.

The most technically challenging requirement was the desire of residents to see a permanently flowing stream in this urban watershed that had lost much of its headwaters. The consultants explained to residents that several times during the year the stream might carry no water because of the altered hydrology of the watershed. City residents and city leaders who had seen renderings of a babbling stream heard this but did not accept this reality.[7] "What sort of stream dries up?" they asked themselves. After some back and forth, Hawley and the team figured out how to make it work. The stream would flow year-round, even if that meant water would need to be recirculated from a downstream pond to the upper reaches via a pump. Some of the more environmentally minded residents suggested that this represented a Disney version of a stream. The majority of city residents and city leaders won out; the stream would have a permanent flow.

When I ask Hawley about the controversy surrounding the pump and other challenges, he framed them as speed bumps along the way. In a critical story, a local journalist said the cost of the recirculation pump could reach $750,000. Although Hawley says he wasn't a personal fan of the pump system, if the city had retained a combined system, an even larger pump would have been required to pump combined sewage to the wastewater treatment plant with a much larger operational cost. That doesn't account for the $200 million that was saved by undertaking this novel sewer separation approach. He mentioned that many affluent communities have expensive fountains, but it became a point of contention in this low-income, predominantly Black community.

Even if Hawley didn't relish these challenges, I came to understand that he saw them as just part of the job. Like Houdini being shackled in a padlocked vest, Hawley saw these multiple and frequently conflicting demands as a problem to be solved. Just as the logs in Shor Park required a certain arrangement, the flowing water necessitated a certain solution. Compared to some of the more firebrand proponents I met at other projects, Hawley's flexibility seemed dispassionately pragmatic.

For this costly and demanding project, Hawley and his partners and the project funders came up with a solution that met most of the demands. Surveying the newly installed project, it's difficult to get hung up on epistemological concerns about whether this is a natural river (figure 4.3). The vegetation, still establishing, grabs your attention with color and structure. The cardinal flower, a plant that seldom survives restoration plantings, is flourishing with its coveted streaks of bright red. The floodplain is carpeted with an array of sedges and perennials that have already become established, their roots anchoring the loose soil that was imported. The channel itself is lined by three layers of tan, stacked stone slabs. Although one might not find a replica in nature, the slabs allow easy access to the stream, and their irregular edges lend a naturalistic feeling. In short, it's attractive, accessible, and unlikely to veer off course.

Outside of the channel, the project provides even more bells and whistles. A large splash park and extensive playground sit fifty feet from the channel.

4.3 Aerial of Lick Run stream daylighting project after major grading but before planting. Courtesy of Human Nature, Inc.

A pavilion and performance space provide ample room for a variety of community activities. Attractive signage describes the historical and environmental context of the project. Waypoint signs direct residents to various sections such as "Headwaters," the "Forebay," the "channel overlook," and "the pond." This signage rapidly places a passerby into the context of this ambitious project. If I lived here, I would be proud of this addition to my neighborhood.

I continue to walk the entire length of the project along a path laid out by the landscape architect (figure 4.4). In a couple of places, the expensive railing that runs along the new sidewalks has been crushed by a wayward vehicle. Random pieces of trash mar the otherwise tidy turf areas and densely planted floodplain. Because of the moderated flow regime, algae has become established in significant quantities in the channel. The overflow system precludes any strong scouring flows that might otherwise flush the algae downstream. Despite the impressive green infrastructure installations that attempt to return a steady base flow, fertilizer runoff from yards that abut them still finds its way into the creek.

4.4 Lick Run planted with overflow grates in the foreground.
Courtesy of Human Nature, Inc.

These cultural and social challenges are difficult to solve in one project, regardless of the project's price tag.

I wonder how the new car shine of the project's amenities will hold up. I wonder if the perennials will develop into a resilient floodplain, and if the new trees will thrive and create the shaded structure important for stream biota. I consider what form this stream, resurrected between two busy roads in a valley that has seen its share of neglect and abuse, will eventually take. Inspired by its 300-year-old unblemished form and cleverly altered to accommodate the reality of twenty-first-century development, it will be something entirely new. An Anthropocene stream in the purest form.

WRESTLING WITH SCALE: CAN THE MICRO-PROJECT BE FRANCHISED?

I return to the unnamed creek late the following morning. Hawley told me that he thought they could finish up today. When I get there, I see that no one is around and all of the log structures are filled with finer branches. A few of the waterlogged locust logs have been cut into shorter sections and deposited in the stream. Having spent a day mucking in this stream, I can say that the site does feel finished. All the logs and branches have found a home. The structures look formidable. I'm wrapping my head around the fact that this project was completed in less than two days at a cost of only $10,000. I receive a text from Hawley telling me they already finished up and to join them at the nearby Bob Evans restaurant.

I hightail it back to the car and onto the asphalt jungle of the interstate exit where the restaurant is located. Hawley and Tower are waiting for their order. I put in a quick order of biscuits and gravy and sweet tea. As we wait, Hawley informs me that Bob Evans started in Ohio and that the restaurant used to be a pretty big deal there. He laughs as he relates that his hometown held two events for their local restaurant; one for the opening and another when the restaurant underwent a major remodeling. I note that his hometown must count a sizable number of appreciative eaters among its ranks.

As I sop up my gravy with an impromptu logjam of buttery biscuits, I realize that I never thought of the restaurant chain originating anywhere. The ubiquity of the chain makes its presence seem almost preordained. Wherever an inexpensive motel chain forlornly sits just off a highway exit, so sits a Bob Evans. It makes me think about what an expansion model of stream restoration might look like. Clearly, with millions of miles of headwater streams whose water quality is trending downward, the market is there. The question is about the demand. Could Hawley's clever method with an attractive price point attract franchisees? Could we as a nation crave clear stable streams as much as pumpkin-spice muffins and pigs in a blanket? Could we redirect Americans' simmering political anger to the industrious, self-determined work that Hawley is offering? Who knows, we just might enjoy it.

CHAPTER 5

BEAVER WRANGLERS

Facilitating Functional River Restoration in Western Washington

HOUSE CALL: UNINVITED GUESTS

We head out to visit a homeowner who had called Molly Alves for help. Camano Island, about an hour north of Seattle, is separated from the mainland by a narrow band of salt marsh braided with tidal channels. Thousands of urban Seattleites have migrated to these idyllic islands in the Puget Sound, and beaver have come along as well, although they are not welcome. Usually thought of as creatures of inland streams and forests, beaver seem out of place on the open coast in this land of mellow rustication. Their obsessive work ethic does not match this place where people come to relax. Beavers are killing the island's vibe.

Alves runs the Tulalip Beaver Project for the Tulalip Tribe. For the last eight years, she has been talking to landowners with unwanted beavers. She first tries to talk them into living with the beavers. If they seem amenable, she'll recommend a visit by a colleague who can help manage the water levels. If the situation seems untenable, she will try trapping them. "In about half of the cases, we can keep the beavers on-site, which is our preference. But if they [landowners] get frustrated, they'll often threaten to call in lethal control and then we'll do it. They kind of have that trump card."

When we get to the destination, the homeowner greets us gratefully. She has owned the home for forty years. She purchased it as a vacation property, but she has lived here full time for the past ten years. The property has increased in value nearly twenty times over the years. She informs me that she loves the location, the expansive view of the sound, and until recently loved the stream behind her house. In the past couple of years, beavers have moved in, probably coming from Puget Sound, which is only fifty yards downstream from the lodge and a network of dams the beavers have built. The creek behind her house, deep in a steep ravine, is now flooded throughout.

The homeowner is worried that the statuesque evergreens growing on the steep hillside behind her house will crash into her dream home and destabilize the bank in the process. The elevated water levels have already contributed to toppling some large cottonwoods on the opposite side of the bank. Fortunately, the cottonwoods fell toward the opposite bank and didn't hit either one of the two towering spruces growing on her side of the bank. She is concerned that the increased water table will kill the spruces and lead to their collapse. She had an arborist come out who confirmed that the spruces would not survive the inundation created by the backwater.

The homeowner takes Alves and her team down to the stream. The dam is not large, but it is creating backwater that floods most of this narrow valley. The beavers have used whatever vegetation they could find to construct it, even some of the prickly blackberry shrubs that cover the hillsides. Alves and her team inspect the dam as well as a drainage pipe that was inserted into the dam in hopes of lowering the water level. She sees this a lot: multiple futile attempts to fight the beavers. Many people physically pull apart the dam, but the dam is quickly rebuilt. People try again, installing drainage pipes only to see them quickly become plugged. By the time they see Alves, they have a healthy admiration for the beaver. This can sometimes work in their favor, especially if Alves assesses that beaver will continue to come to the site. The homeowners, having fought for years, are war-weary and open to a negotiated settlement. A properly installed flow device that keeps water levels manageable can be the linchpin that leads to an armistice between the opposing keystone species. Humbled, the property owners are often willing to pay for the service.

But in this case, Alves is open to removal. The immediate threat to the landowner's home is clear, and Elyssa Kerr, Alves's colleague and a flow device expert, didn't recommend such a device after visiting the site. Given the unstable hillside and the newly saturated conditions, it would be unwise to suggest anything that could threaten people's homes. These beavers will soon find a new alpine home in the Cascades and continue their focused work unbothered.

THE UNLIKELY TRAPPER

It's late in the afternoon and Dylan Collins is inspecting the area where the trap will be placed. A wildlife technician who works for Alves, he stalks around the area with cougar-like caution, taking care not to disturb too much of the vegetation. He points out a row of sticks that extend outward from the area of interest; he calls them guide sticks. He put them there earlier to coax the beaver to approach the trap head on. If the beaver come to the trap from this preferred direction, the trap will snap together and sandwich the rodent neatly between the two sides of the trap. Collins's arms are a tattooed menagerie: an alligator and an eagle sit next to a mountain range, a skull reminiscent of *T. rex* bumps up against a lotus-worshiping puma. Given his gentle demeanor, I rule out some sort of animistic cult. I ask if these images are a pictorial resume of his trapping. Beavers are the only animals he has trapped he tells me.

I notice only the most obvious beaver indicators. A six-foot-high lodge sits on the edge of the pond, placed right above the inflow pipe that drains a twenty-year-old subdivision across the street. Only about fifteen miles from downtown Seattle, the landscape retains much of the verdancy of the original forest. Ponds and flowing creeks are inconspicuously linked as they flow through wood lots and yards planted with a mix of ornamental and native trees. The jagged peaks of the Cascades lie just thirty miles to the east. It seems like a nice place to live. There's also a lot to like if you're a beaver.

King County wanted the beavers removed to ensure that the stormwater pond would work as intended. In some places near Portland, they are experimenting with letting the beavers stay. But that's Portland. Nearly everywhere else the beaver are unwelcome. Yet to the beaver we have provided a pond and flowing water in the form of inflow and outflow pipes. Beavers are experts at making the best of what is offered. Stormwater engineers didn't anticipate this when drawing up their plans in AutoCAD. It's as if the beaver are saying, "We see what you're trying to do, but let us make it better."

Humans spend millions to construct and maintain these artificial ponds. Like a vagrant nephew who mows the lawn in exchange for room and board, the beaver don't need to be paid. Their requirements are minimal: flowing water running at less than a 6 percent slope and an adequate supply of woody material.

Aspen, willow, poplar, birch, and cottonwood are preferred. Sedges and black-berry canes will work in a pinch.

Collins is back in the pond. He checks the depth of the water. Along the edge of the pond, swaths of rushes extend landward from a surprisingly placid and natural looking stormwater pond. Birds flit around the edges unconcerned with our activities. The pond has some deeper spots, but Collins has located the two traps far enough away from those areas so that a trapped beaver couldn't drag the trap into the depths and drown. He places the bottom of the trap in shallow water at the edge of the pond and checks the trigger mechanisms. Then he begins the baiting.

Baiting in this context is done with scents. He opens a black plastic repurposed ammo case that contains a dozen different glass vials: timber, bread and butter, woodchipper, backbreaker (figure 5.1). Self-conscious about the aggressive tenor

5.1 Dylan Collins shows off his potpourri of scents intended to lure beavers into his live traps.

of the names, he mentions that most of these are marketed to kill trappers. We open some and smell them. The first one is reminiscent of root beer. The next is also strangely pleasant, but the odor is unrecognizable. One has a pungent castor smell whose scent lingers, not in a pleasant way. Collins baits one of the traps and leaves the other for two colleagues who are helping him today. He asks them what they used last time and offers a gentle suggestion for using a little woodchipper with something else. It's a subtle nod to artistic license in an otherwise highly technical process.

Collins tells me that he's talked to some trappers about other animals. Apparently beaver and coyote are the hardest mammals to trap. One trapper described how he caught a coyote by intentionally placing a tennis ball along a trail. Only after the ball was removed, three days later, did the coyote come to inspect the site where the tennis ball had been. That's when he set his trap, and that's how he caught him. "Some of these trappers are more knowledgeable than any wildlife biologist."

The scent is placed on a foot-long branch of willow weaved through the side of the trap farthest away from the pond. Collins runs a small amount of back-breaker and woodchipper along the leaves. They've been here for more than a week trying to get this last male. With the mother and two kits already secure at the hatchery and ready for deployment, they'd like to get this last one and move on. Lots of people are asking for their services, and the trapping season is short. They trap only in the early summer after the kits have been born. The goal is to keep the family intact and release them early enough in the season so they can become established before winter.

I ask Collins how things will go down. When the trigger is released, he tells me, the spring will instantly lift the half lying in the water and throw the beaver into the other half. Sandwiched like a burrito, the beaver will have limited mobility. The confinement shouldn't be overly stressful, Collins says, because beaver are accustomed to being in tight spaces. Collins will return around seven in the morning to check the traps. They don't want the beaver confined longer than it needs to be, and they'd rather avoid the attention of local residents. Although they are focused on trapping in the most humane way possible, any trapping of wildlife is sure to bring unwanted attention. Humans don't want to be reminded that our created landscape seldom allows these animals to coexist with us.

TAYLOR CREEK: BEAVERS DELIVERING FOR
THE ENDANGERED COHO SALMON

Three six-foot-wide banners exclaiming "Jesus King of Kings, Lord of Lords," "Yeshua," and "Jesus said, Repent" occupy the central spot in an otherwise unremarkable neatly mowed yard. Next door is a vacant lot fenced off and locked. Tall grasses grow through a gravel access road. This lot serves as an access point to a larger section of public land that has been used for wetland and stream mitigation projects in the past twenty years. The fervor of the signs contrast with the far less heralded return of the beaver and the ecological salvation they quietly deliver. Instead of shining a light on those who would not be saved, the beaver relentlessly work to summon additional water into this floodplain to the benefit of the endangered Coho salmon. Yet the raised water levels would threaten the neighbors' outbuildings. This situation had none of the finality presaged by the signs. It is a situation with an uncertain outcome; judgment day is postponed while opinions are being shaped and formed.

We are only two hundred yards off Highway 169 in southeastern King County in the floodplain of the Cedar River watershed. Despite its proximity to the busy road, the area feels rural and removed. We pass through narrow, timber-supported overpasses that hold up an older, unused rail bed. Ahead of us is Taylor Creek, an environment altered, I would learn, by ongoing restoration work. The area is low, the air moist, and the atmosphere heavy on this sunny morning. The surrounding homesteads are large and of varying upkeep. Some folks have free ranging chickens, some have tidy lawns. The verdant vegetation on public land on the far side of the stream towers over the homes and waits in leafy expectation. The recovering forest seems intent on returning to the temperate rainforest it once was. If left alone for five years, it might do just that—recolonizing the turfgrass, overtopping roofs, and providing the free-ranging chickens with an even wilder pedigree.

Jennifer Vanderhoof, senior ecologist for King County, organized this site visit for me. We met at BeaverCon about two months ago where she presented her "Planning for Beavers Manual," which she had been working on for more than a year and had hoped would be released to the public. Her official title didn't immediately point to her expertise and training in wildlife biology. Originally from Missouri, her first gig out west was monitoring fishing boats in Alaska. After Alaska and a short stint with the city of Seattle, she took a job with the county. As an ecologist, she's been pulled into various projects over the years. Much of her work in the past five years has been related to wildlife, and most was related to beaver. Although her smile gives this away, she adds that this recent stretch has been the most fulfilling in her thirty-year stint with the county.

At the conference, I spotted her convening with other people from the northwest who all seemed to know one another. As I would learn, she is part of a trio of women who handle most beaver conflicts in the counties surrounding Seattle. Vanderhoof is the county government representative called in to advise when beavers invade a stream-restoration project or a stormwater structure. Alves, an employee of the Tulalip Tribe, has taken on the role of trapping and relocating problem beavers; and Elyssa Kerr, director of Beavers Northwest, specializes in flow devices and other contraptions that regulate water levels in beaver dams. They work closely together because many sites eventually require their attention at varying points in the process.

Humans seem to be coexisting with beavers at Taylor Creek, but that coexistence is fraught. The riparian area is the site of several ecologically related wetland and stream-restoration projects overseen by King County that began in 2001. Most of the projects were funded through fee-in-lieu programs that developers paid into for the right to destroy wetlands elsewhere.

The site was chosen for its potential to improve habitat for the Coho salmon. A secondary issue was flooding that had plagued the area. Vanderhoof explains that the stream regularly flooded the adjacent road, usually along the edges but occasionally covering the entire roadway after heavy rains. One of the early projects removed a culverted section that ran behind the home with the evangelical banners. Another early project added foot-high berms next to the roadway to keep water from flooding onto the road.

As we walk up to Taylor Creek, we pass through a fence with an odd opening. We weave around a second three-foot-long fence that runs parallel to the main one and partially blocks the opening. The project manager for the site says that this sort of arrangement may deter beavers from dragging trees toward the stream, but he doesn't seem fully convinced. Although beavers have colonized a wetland area adjacent to the creek, he is hoping they don't set up shop directly on this section that runs behind the home with the religious signs. Even without a dam in the main channel, their backyard and the small barn located close to the stream are regularly inundated by high water. According to the project manager, the flooding doesn't reach biblical proportions, but he's not so sure the property owners feel the same way. He has met with them about the issue and found that they weren't particularly opposed to the beavers. He also mentions that the most uncomfortable part of the meeting was when they asked for his thoughts about the afterlife. "I'm recently divorced; I'm certainly going to hell," he says.

Although the evangelical overtone is unique in this case, Vanderhoof says that they see a lot of these situations. They hope to move barns away from the stream at no cost to the property owner so impacts from the beavers will be less of an issue. A couple of years ago they had the full support of a King County council-member who championed the program. The representative cheekily named the program "Rebarn Again," hoping to appeal to the humor of rural folks. In this case, the name backfired. The religious family next door took personal offense and have since indicated that they're not interested in moving their barn. Currently, they aren't complaining about flooding, but the project manager and Vanderhoof worry that the issue isn't going away. The project manager testifies that the newly created wetland and stream complex is paradise for Coho smolts, but the issue with the roadway flooding remains much the same, if not worse, as it was twenty years ago.

This situation isn't unique. It's also why Vanderhoof has been pushing to release her manual. Its currently the middle of July, and the manual still resides in administrative purgatory. She has been assigned a technical editor but has been informed by her boss that "this one should go all the way up the chain." I try to understand the level of clearance required to access the document, but apparently this manual has some people worried about its implications. When I ask her to describe what guidance the document will offer project managers, she says, "first, ask what the potential backwater impacts might be, and second, ask what potential impacts to vegetation, existing or planted, can be expected." This foresight doesn't seem dangerous enough to warrant the document's sequestration.

Despite of the bureaucratic limbo of the manual, she's regularly engaged in these situations where beaver have become "issues." These issues have resulted in costly modifications to restoration projects, some of them becoming endless money pits. If they had known what was coming, she says, they probably would not have undertaken the project. She tells her fellow colleagues, "If you plant trees, expect beaver to show up." King County has hundreds of miles of ideal stream habitat with low gradient streams and abundant vegetation. All the restoration project managers she meets with are now thinking about beavers. They may not know where the beavers will build, but they're thinking about them. "That's a change over the last ten years," she tells me.

It's an uncomfortable and unspoken reality in government that there is little love for planning documents or manuals. Those who are familiar with the issues believe they already know what will be contained in these documents. For them, manuals merely act as a backstop to formally document what they have already been doing. They might support the development, but they aren't going to use the manual. For others, these planning manuals loom large in their potential reach. Suggestions contained therein can be applied to any number of potential scenarios, many accompanied by some negative attention from media. Then there are vested interests, the sister agencies or any number of people who might find the manual constraining. One can hear the teeth gnashing in the corner offices of upper management. For restorationists, planning for a species that is likely to enter your project, alter the work, and transform the landscape is more than good sense. It is central to predicting the end result of this work.

Although Vanderhoof appears excited about her current work, signs of frustration built up over a thirty-year career in government occasionally slip through. Some of the greatest barriers to human-beaver coexistence stem from her own agency. The Farmland Preservation Program, also administered by King County, buys the development rights from farmers to preserve their lands as farmland. Where these farmlands abut streams, it would seem that the soggier floodplain beavers precipitate could be accommodated. Unfortunately for the beavers, these programs stipulate that 95 percent of the farmland must remain in production. Similarly, salmon protection efforts such as stream buffer plantings have led to situations in which the goal of a healthy stream stands in direct conflict with the goal of preserving farmland. "First you plant your buffer, then beaver come in and expand your riparian area," Vanderhoof tells me with a sigh that suggests she's often witnessed this outcome. The inevitability of this outcome, however, has not yet resulted in any policy modifications in these farmland programs. Apparently not everyone is thinking about beavers.

We walk through the restored area, passing the first site completed in the early 2000s. The trees are fifty feet tall after only twenty years. The area is always wet, but this year it has reached record levels. They've had significant rainfall

every week since early January. As we walk through the stream and wetland complex, the project manager mentions that the Coho smolts reside here year-round,
unlike in other coastal environments. The area is fed by many cold springs that
keep the temperature cool in the summer. Vanderhoof points out beaver browse,
affectionately known as "chew," in numerous spots. A complex of beaver dams
has raised the water level two feet above the prerestoration condition. This was
modeled before restoration, the project manager tells me, but they both seem
surprised by the expanse of water.

The high water has created large ponds that are mostly unshaded by the
surrounding trees. The project manager wishes they had created some higher
islands in the center that might support trees to provide additional shade for
the Coho salmon. Vanderhoof points out a cottonwood, blown over by a storm,
that stretches halfway out into a pond. Several branches from the tree are now
shooting up vertically, creating a row of new trees that promise shade in the near
future. "This sort of thing is in the manual," she says. "I get a lot of ideas from
project managers and include some of my own." In this instance, the wind did
the work, but she suggests laying these downed trees in water, either felling them
by chainsaw or placing them by helicopter. The downed trees can provide shade,
and the multiple stems can better withstand herbivory.

We head through this soggy water world to a site where there's been active
beaver logging. Several stumps, from four inches to eighteen inches in diameter,
display the telltale chew. Vanderhoof raises the question of whether beaver intentionally direct the logs toward the water. It would make sense for the beaver, but
how smart are these rodents really? One would have to factor in the slope of the
land adjacent to the pond, which would predispose any felled tree toward the
pond. As if to further pique our interest, one twenty-inch tree still stands defi-
antly, despite being chewed to a narrow hourglass shape two feet off the ground.
We inspect this tapered section of the trunk and look for any signs of intention-
ality. We try to get inside this beaver's head. Time slows as if we're watching
grains of sand flow through an actual hourglass timer. The beaver has chewed
away more from the pond side of the tree, but it's hard to tell how that might
affect its future flight path. Even when accompanied by experts, I come across
many of these "how do they do this . . ." moments. It seems a fitting question both
for serious scientific inquiry and for circular substance-influenced dorm room
discussions. In addition to eliciting respect for their work ethic, beavers also
seem to provide some sort of levity.

I wonder if the research and accumulated knowledge of these professionals will lead to greater acceptance of the beaver. Some of our aversion to this creature seems to be based on its unpredictability. Plugging the tasks of a beaver family into a Microsoft project time line is difficult. We'd have to assume that some of their impacts will be an impingement on our way of life but that broader benefits will outweigh these impacts. We'd have to accept that messy streams are healthy streams. We'd have to accept that beaver aren't varmints to be killed but creatures intent on building the aquatic environments we profess to love. It requires faith in the hard-wired instincts of these creatures. Their work might be more ecologically beneficial than what the accumulated knowledge of humans in multiple disciplines has cooked up thus far. It requires a humility uncommon to our species.

ACCEPTING UNCERTAINTY IN STREAM RESTORATION

Stream-restoration practitioners attempt to understand the uncertainties that can affect a restoration project in many ways. Fluvial geomorphologists and engineers who model the flows under different channel configurations attempt to understand sheer stresses under different flows and how those will affect whatever material is placed on a stream bank. Ecologists and landscape architects use a knowledge of plant growth to anticipate how specific plants will colonize an area and how the growth of those plants will create a new canopy that can both shade the stream and provide stability to the banks through the network of roots. Fisheries biologists look at different stream features, pools, riffles, and runs, as well as macroinvertebrate populations that sustain fish. Those studying water quality look at how the connection to the floodplain can boost capture of sediment under flooding conditions and augment the processing of nutrients.

These predictions, some mathematical and others based on a deep understanding of the interactions between species, all assume that the channel created during a restoration project will continue on in that fundamental form. Just as a home builder would expect the walls to remain upright, those restoring streams assume that the stream will remain distinct from a lake, a wetland, or a bayou. Adjustments are expected; the stream will form new sediment bars in certain spots, and in other spots it will scour out pools. But overall a stream is expected to flow downhill in a single channel along a relatively sinuous path. After all, that is what a stream is.

In an increasing number of streams, however, beaver upend this understanding by bringing major changes to the fundamental structure of the channel. Damming changes the passage of sediment and nutrients downstream and modifies the movement of fish in both directions, sometimes termed vertical connectivity. If the dam remains in place, the primary single-threaded nature of the channel will ultimately change to something braided or anastomosed as the stream begins to find new channels in the floodplain to which it has now gained access. What was a stream now more closely resembles a wetland or a bayou. When a creature arrives whose life cycle depends on the consumption of vast amounts of plant material, the prior thinking on tree and shrub planting evolves into a process of triage, mitigating damage while trying to accommodate the new arrival's voracious appetite for woody material.

For restorationists, this situation can be both maddening and humbling. Practitioners spend hours carefully designing projects that will eventually deliver the ecological outcomes that are promised. Sometimes boiled down to the phrase "ecological uplift," this overarching goal ultimately serves as the litmus test of the entire endeavor—reflecting the knowledge, training, and experience of all involved. When beavers add their own touches to a restoration project, it nearly always dramatically alters the appearance and function of the project. After the beavers have had time to establish, the ecological benefits of their work oftentimes exceed what could have been done without their involvement. The situation brings with it an unnerving existential quandary. How do you justify your specific expertise when a large rodent that has occupied most of continental North America does a better job at no cost to the client? As one prominent consultant half-jokingly said at the BeaverCon conference, "there are a lot of well-constructed cross-vanes in this project buried under those beaver dams."

Amazingly, many stream-restoration practitioners are taking this in stride. One nexus of activity is in western Washington where beavers are attracted to the low-gradient streams at the base of the Cascades. A few individuals are working as close to the front lines as one could in this field. They are identifying, tracking, and relocating beaver. When it is possible, they modify beaver dams to find an acceptable balance with this species that is acting as an unpaid consultant writ large. They are communicating with the public that mostly professes love for beavers but can't tolerate what beavers do. They are talking to colleagues who manage restoration projects, urging everyone to anticipate the arrival of beavers and to create restoration projects that will evolve with their occupation. When the beavers must be moved, they are relocated in the forested western slope of the Cascades where they can implement their ecological restoration in peace. This type of restoration requires a level of humility and acceptance of uncertainty that has been absent in stream restoration. It also requires a high degree of coordination and collaboration, evident in the work of these three women. In a field that rewards confidence and authority, they are betting on an industrious rodent. Assessing this forest of human-derived solutions, the beaver goes about felling each one, putting it to another use, unplanned and unimagined, but undeniably beneficial.

———◆———

We met at the Safeway in North Bend, Washington, at 10:30 a.m. sharp. Alves wants to get the beavers into the stream before the heat of the day permeates the covered bed of the truck. After a quick greeting and rundown, her three employees take the truck with the beavers, and Alves, Vanderhoof, and I follow in Vanderhoof's car.

In her early thirties, Alves tucks her dark hair in a ponytail and wears a T-shirt and yoga exercise pants that work well with the waders she puts on several times a day. We have been conversing for more than a year and have met at BeaverCon. She has been expecting me and has managed to fit nearly all aspects of her job into my two-day visit. The recent trapping of a small family of three beavers lines up perfectly with my visit. Vanderhoof found them living inside a storm grate that covered the outflow pipe in a stormwater pond in Renton. I'm flattered by the intense scheduling, mostly for my benefit, but learn that Alves has become quite adept at dealing with media. She was recently featured in the *Wildlife Nation* TV show, is the subject of an upcoming article in *National Geographic*, and has been approached by BBC, among others. "It's probably 40 percent of my job now; but it turns out I'm pretty good at it, and it's important for what we're doing."

I probe this subject warily, realizing that I am a small squall in this larger media storm. Alves says she is as accommodating as she can be, but ultimately we're on the beavers' schedule. A BBC reporter wanted to visit in January, and she reminded the reporter that the concrete sluiceways in the tribal hatchery would be empty then. "There's definitely a cuddly angle to the story that is irresistible." I cringe as I think I may be falling for cuddly as well. Vanderhoof has received a lot of interest from other media outlets as well. Talking to Alves she says, "Remember that *Grist* reporter that made me sit in a chair for two hours. He only used a couple three-second disjointed snippets. I think they only included it because they made me sit there so long."

We are heading into the Snoqualmie National Forest, the eventual location for these lucky beavers. Vanderhoof and her team have been relocating beavers in the Snohomish River watershed immediately north of this one. The U.S. Forest Service has been so impressed with the results that they asked that beavers be released down here as well. In the headwater areas where beavers have been released, they have quickly gotten to work. They have added significant wetland habitat, which in the Snohomish has provided additional salmon-rearing habitat. The expanded riparian areas also provide a much-needed firebreak. Given the rash of wildfires in the western United States, this benefit is of particular interest.

BEAVER BENEFITS: A CHANGING AND DYNAMIC FIELD OF SCIENTIFIC INQUIRY

The work of Alves and Vanderhoof represents the front line of beaver management, but many others are working to understand the benefits beavers provide. Others are putting in a good word for beavers in the course of their regular work, advocating for "nature-based" solutions. Still others incorporate this work in their restoration business model. The activity spans many climatic zones, incorporates different skill sets, and is adapting to a changing public perception about these rodents.

This openness toward the aquatic ecosystem engineer was not always the case. For most of the twentieth century, beavers were seen as impediments to effective stream restoration. A publication produced by the Wisconsin Department of Natural Resources (WDNR) in 1983 titled "A Bibliography of Beaver, Trout, Wildlife, and Forest Relationships" is notable because it reflects the perspective of people most responsible for deciding the beaver's fate. It speaks candidly about the dominant opinion of beavers at that time:

> The degradation of trout habitat by beaver is considered a severe threat to wild trout management in Wisconsin. Beaver impoundments serve as heat collecting units in summer and cold storage units in winter, markedly affecting survival of resident trout. Such impoundments often lead to the buildup of fish that compete with trout for food and space (e.g., suckers, chubs, daces and shiners). Higher fish densities in the impounded areas also attract greater numbers of avian and mammalian predators and increase the likelihood of fish parasites and disease. Decreased water velocities and the sloughing of inundated stream banks within impounded stream segments contribute to the siltation of former gravel riffles used for food production and trout spawning.

The publication goes on to highlight timber losses caused by beaver dam–induced flooding and the "direct cutting of 3,000–4,000 acres of aspen." Even when the beavers are gone, the remaining dams are implicated in continued problems. "Stream velocities are not sufficient to cut through silt deposits and re-establish a pool-riffle complex similar to that present before impoundment. Stream water temperatures remain higher than before impoundment because of the wider stream channel and lack of forest canopy for shade." This thinking led resource managers to remove thousands of beaver dams in relatively untouched stream and rivers in the upper Midwest.

Current researchers tell a dramatically different story. Emily Fairfax has focused on the benefits of beavers in a landscape after major forest fires. There is a general understanding that beavers create a wider, wetted floodplain but exactly how that soggy area fares during and after increasingly common wildfires is not known. As Fairfax states, "When it comes to water, beavers slow it, spread it, store it."[1]

Fairfax and her team used the satellite-derived Normalized Difference Vegetation Index to determine greenness of vegetation from remotely sourced aerial images. She looked at "data from the year before, year of, and year after a major wildfire to examine the vegetation changes of riparian corridors with and without beaver." The riparian corridors were subject to forest fires of varying intensity. Although the exercise was primarily conducted in a lab using aerial imagery,

Fairfax describes going to some of the sites firsthand. At BeaverCon, someone asked if beaver-influenced riparian corridors acted as a firebreak. (Not that she was aware of, but this doesn't mean that it wouldn't be a possibility.) Riparian areas are commonly used as a starting point for a backfire that enables firefighters to contain forest fires. Certainly it would be asking too much of these rodents to stop fires. But what they leave has cascading benefits that researchers are just beginning to quantify.

One line of study involves understanding the degree to which beaver dams can slow the massive export of sediment, ash, and woody debris downstream following a fire. Ellen Wohl, a noted river geomorphologist working out of Colorado State University, conducts research that attempts to quantify the benefit. It is well known that denuded landscapes following forest fires lead to changes in stream hydrology. Moderate rainfall events that were previously intercepted by forest canopy and infiltrated into the soil now course downstream and create flooding conditions. These events carry massive amounts of soil from slopes no longer stabilized by trees and shrubs. Charred trees and other downed debris are also flushed downstream where they frequently clog and ultimately destroy culverts. Wohl sums up the scenario this way: "As a warming climate drives more intense and frequent wildfires in drylands around the world, it becomes increasingly important to understand the characteristics that foster resilience to the wildfire disturbance cascade."[2]

Fortunately, the types of streams beavers tend to occupy, smaller to medium-sized first to third order streams, comprise the majority of stream miles in forested watersheds. Their dams can detain the downed and charred wood as well as vast amounts of sediment. Contrary to the denunciation laid out in the WDNR document, dams even provide benefits when the beavers abandon them. Wohl describes a process whereby abandoned dams "attenuate post-fire sediment inputs moving down the channel."[3] This sediment is stored behind dams and settles onto floodplains created by these dams. The early successional vegetation that colonizes this new growing area is often the type that is favored by beavers. The old dams create the condition for new beavers to return to the site.

In much of this research related to the benefits of beaver, there are concerted attempts to describe the multiple, interconnected benefits of these beaver meadows. The scientific description of these areas speaks of anastomosed channels, secondary flooded channels, expanded hyporheic zones, and increased utilization of stream carbon. Between the lines of the constrained language of academic articles, one can hear exuberant joy in the benefits these creatures provide. One can imagine that beavers are somehow intuiting the entire range of stream processes that researchers care about and are quietly going about facilitating them. Yet the way they go about this, quietly and randomly, makes the quantification and explanation of these benefits challenging.

An influential 2014 article in *Bioscience* written by Michael Pollock presents graphics that help to explain the alchemy beavers initiate.[4] The forlorn incised channel, with its steep banks, narrow channel, and rigidly linear flow transforms with the addition of beaver dams. Sinuosity returns when the beaver dams are outflanked by blowouts or end cuts. The resultant erosion of the steep banks leads to additional sediment that piles up behind downstream dams. As the base elevation rises, the stream forges new side channels that weave through this new stream base. Logs and other debris are captured behind old dams, and vegetation eventually reemerges in the floodplain. The beavers, in their single-mindedness, create new dams that continue this process of turning what was a simple single-threaded channel into a complex, multithreaded wetland (figure 5.2). To anyone versed in ecological processes, this evolving assemblage of water, wood, vegetation, and beavers is not just an abstract process—it is the embodiment of restoration.

Although ecologically satisfying, this process cannot happen in all incised and degraded stream channels. Particularly in the arid west, where riparian vegetation is limited due to the commonly diminished floodplains, there simply aren't enough trees and shrubs to sustain beavers. In other cases, the flood flows quickly wash out beaver dams. Understanding the benefits that beavers provide, and seeing streams that desperately needed these benefits, some people started to build a human constructed version. Known as beaver dam analogues (BDAs), these structures attempt to jumpstart the process. Typically, a series of logs are driven vertically across the channel to provide structural support. Next, logs are placed horizontally to begin backing up the water. In most cases, these BDAs are installed in areas close to beaver habitat. The hope is that a few of these BDAs will create an expanded floodplain and additional floodplain vegetation. Beavers might take over the BDAs or create their own dams. Either way, in deeply incised streams, the reinforced BDAs typically fare better during flooding events. In this perhaps most experimental approach toward stream restoration, the technical specs are taken directly from the work of a rodent whose purely instinctual approach makes citation challenging.

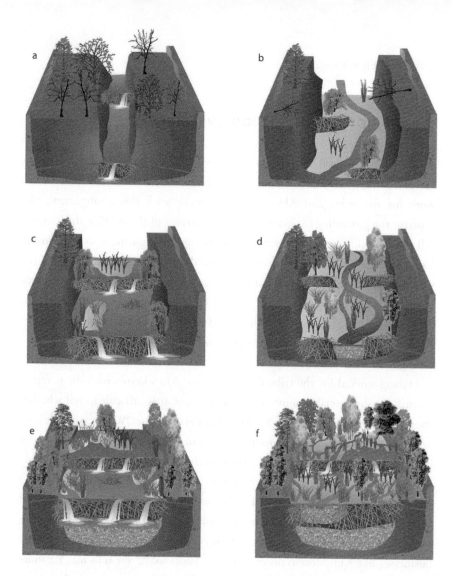

5.2 How beaver dams affect the development of incised streams: (*A*) Beaver will dam streams within narrow incision trenches during low flows, but stream power is often too high, which results in blowouts or end cuts. (*B*) This helps widen the incision trench, which allows an inset floodplain to form. (*C*) The widened incision trench results in lower stream power, which enables beavers to build wider, more stable dams. (*D*) Because recently incised streams often have high sediment loads, the beaver ponds fill rapidly with sediment and are temporarily abandoned, but the accumulated sediment provides good establishment sites for riparian vegetation. (*E*) This process repeats itself until the beaver dams raise the water table sufficiently to reconnect the stream with its former floodplain. (*F*) Vegetation and sediment eventually fill the ponds, and the stream ecosystem develops a high level of complexity as beaver dams, live vegetation, and dead wood slow the flow of water and raise groundwater levels, allowing multithread channels to form that are often connected to off-channel wetlands so the entire valley bottom is saturated.

Michael M. Pollock, Timothy J. Beechie, Joseph M. Wheaton, Chris E. Jordan, Nick Bouwes, Nicholas Weber, and Carol Volk, "Using Beaver Dams to Restore Incised Stream Ecosystems," *BioScience* 64, no. 4 (April 2014): 279–290, https://doi.org/10.1093/biosci/biu036.

TREATY RIGHTS AND TENSION OVER ACCESS

As the road turns noticeably from paved to deeply potholed, I ask Alves about her role with the tribe. It is increasingly common for nontribal members to work for the tribe, and Alves appears content with this arrangement. She explains the grounding for her work, the 1855 treaty of Point Elliot that ensures full participation in management of the fishing and hunting resources on the Usual and Accustomed grounds. For the Tulalip, that means a majority of the western slope of the Cascade Mountains. Her work applies those rights in a way that references ecology more than the law. By saving and relocating the beaver, she is allowing the beaver to establish ponds that will sustain a range of species. Rather than litigation that extracts punitive damages, this solution offers cascading benefits to all.

Having worked for the tribe for eight years, Alves knows who she is representing. As she describes some contentious issues, it is difficult to tell whether she is describing the tribal perspective or her personal one. "We occasionally butt up against the state and user groups that influence them." Recreational fishing interests have the ear of the DNR. The fees from fishing licenses support many of the staff that manage these programs for the state. Even hiking access can be a contentious issue. "The majority of recreation infrastructure enhancements and additions are being funded and carried out by organizations like the Mountains to Sound Greenway Trust that get millions of dollars from interest groups and businesses like REI. More trails equals more boots sold!," she tells me. Yet more hikers means fewer harvesting opportunities for tribal members.

The massive potholes in the dirt road slow Alves's truck to a crawl. There is plenty of time to think and to talk. Initially I find it difficult to fault hikers clad in expensive Patagonia gear walking through the woods. After another fifteen minutes of chatting in which we cover only a few miles, it becomes clear that this access isn't always benign. Alves doesn't come across as strident, rather just clear in her perspective. I think of Edward Abbey's polemic on Industrial Tourism and see a direct line from his incendiary warnings about paving roads in the National Parks of the southwest to Alves's concern about the ever-expanding demand for access to the forests surrounding Seattle. If you've spent hundreds of hours trapping and moving beavers from habitat that has been dominated by humans, the introduction of humans to wild areas doesn't seem to be a positive development.

I see her and the tribe's perspective more clearly later when I examine the Mountains to Sound Greenway Trust webpage dedicated to the Middle Fork Snoqualmie Valley, the watershed where we released the beavers. "Just 36 miles from downtown Seattle, this wilderness playground defies comparison. No other major population center stands so close to such vast and accessible public lands."[5] The webpage extols the virtues of a region extending from Ellensburg to Seattle. Maps of the area are alluring and invite exploration. The expansive goals of this sort of landscape level effort, calling the four million residents of greater metropolitan Seattle to go forth and recreate, seem less wholesome, especially from the perspective of tribes who have been placed on discrete parcels of land.

For the tribe, this expanded access means fewer hunting opportunities. Even if the wildlife are not driven into hiding, the presence of thousands of amateur outdoorsmen is likely to dampen one's enthusiasm for hunting. When hunting would be welcome, as is the case with damage permits for elk, the tribes aren't regularly consulted. When elk damage a farmer's crops, the DNR frequently issues a nuisance permit to kill the offending elk. Farmers get rid of the offender and can keep the elk. Alves maintains that these animals should go to tribal members. Elk hold special significance to the tribe as a traditional food source, and the coastal 35.3 square mile reservation cannot sustain a herd, she explains.

We finally reach the designated beaver release point. As if waiting for an unfamiliar ritual to begin, we form a semicircle around the back of the covered bed of the pickup truck. The beavers are inside, partially covered with a blanket. I'm expecting some sort of dramatic movement when we uncover them. But when revealed, the beavers just sit quietly in two large Havaheart rectangular traps with sliding gates at both sides. The traps are of the same style as the live traps for mice but decidedly larger. A mother and one kit huddle in one trap, and another kit sits quietly in the other (figure 5.3). Vanderhoof heads to a location downstream where she plans to capture the release with her camera. Moving purposefully, Alves, Collins, and two other techs lift the cages out of the pickup truck and take the beavers down to the shallow stream. They place the beavers in two inches of clear flowing water as we stand around watching them.

The beavers seem unfazed. While they are just sitting there, I take the chance to get a closer look. I realize that I've never seen one so close. They are undeniably cute. "I call the babies furry potatoes," Alves says. She's seen this reaction before.

5.3 A young beaver kit is awaiting release into a mountain stream in the Cascades in Washington.

The sort of quiet satisfaction of seeing an animal in its ideal habitat. "Do you want to do the honors?" she asks me. I'm happy to oblige, and we simultaneously open the gates as if we were at Churchill Downs (figure 5.4). The beavers don't react on our schedule. The kit in my trap looks back at me as if looking for assurance or wondering if I will follow. After a minute goes by he ambles out and starts floating downstream toward a large pool in the stream. The mother in the other trap heads out as well. The remaining kit in the other trap remains. We all watch silently. A few minutes later he violently slaps his tail but still doesn't leave. Alves lifts one end up and nudges him out. He eventually slides out and head downstream toward the others.

Vanderhoof rejoins us beaming, "great footage—I got it all." Earlier, in the Safeway parking lot, she had alluded to the fact that there had been four beaver at the stormwater pond. "I don't really want to talk about it," she confides. Shaking her head, she elaborated, "the worst thing is that it was my fault." Realizing that she was wracked with guilt and was beating herself up about it, I let it go.

5.4 Molly Alves and her team release a family of beavers into a tributary of the Middle Fork of the Snoqualmie River.

Now she is basking in the warm feelings of the moment. It's hard not to feel good about this work, and I sense that Vanderhoof needed this. This important supporting role in a yet unfinished stream-restoration feature film somehow steals the scene. The refreshing level of foresight and planning delivers an Oscar-worthy performance. Placing these purposeful rodents where they can fulfill their role in restoring streams leaves me feeling buoyed and imperceptibly connected to a larger ecological process. Through a web of interspecies relationships, we can gnaw and tug our rivers back to health.

IF A BEAVER BUILDS A DAM IN THE WOODS, IS THE STREAM RESTORED?

I asked to see some of the areas where beavers from previous releases had become established. Alves has taken us a mile upstream from the release site. Here the channel is wide and full of clear pools, and emergent vegetation pokes up from the edges. Alves leads the way, her team following, and I a bit further behind navigate the unknown depths in my sandals. The water is bracingly cold, and the sediment that squishes through my toes is a tactile reminder of the benefit of these dams.

When we come to a dam, Alves reminds us to walk gently over the top. The sticks are embedded in a mixture of mud that settles and adjusts as it accommodates our weight. This dam, the direct product of the beavers, is also the indirect result of a lot of trapping and relocation. Alves, the ecological engineer once removed, appears satisfied. We pass a couple of camera traps, and Alves tells me that these traps are the primary source of information about the results of their work. They see beavers, but they also see lots of other wildlife, including bears and mountain goats. Occasionally the cameras record people posing for selfies.

We're quiet as we wander upstream through the pools. Moss-covered logs extend over the stream. Backwater channels extend watery fingers in unpredictable directions, inviting aquatic macroinvertebrates and juvenile salmon. Green leaves recently shed from streamside alders float on the surface, and submerged brown ones shed at some earlier time point to the carbon capture that is taking place. As if in some sort of riverine church, none of us talks much. The ecological fullness of this place appears self-evident. It needs no signs proclaiming its glory.

BIG-TECH BEAVER AND THE PROMISING FUTURE FOR A BEAVER ADVOCATE

We're at an ice cream place in North Bend getting milkshakes. This is part of their celebratory routine after releasing a new family, Alves tells me. North Bend is full of new buildings and newly constructed roundabouts that keep traffic moving. At the sides of the new intersections, bioretention cells have been installed to capture runoff, but these structures do not allow water to stand more than forty-eight hours. The beavers will not be tempted.

I ask Alves about her future plans. She tells me she is heading off to a graduate program at Utah State this fall. Utah State has become a hub for scientists researching process-based restoration. These researchers have engaged in numerous projects in which beavers have been reintroduced to incised and beleaguered western streams. When the beavers have had sufficient food sources, there have been dramatic changes in the riparian corridors. Once narrow and ecologically depauperate conditions have, in a few years, changed into wide, luxuriant floodplains. In the arid west, these reborn floodplains act as refuge for a wide range of mammals and provide the conditions for improved fisheries.

Alves's Google-funded study is a strange but fortuitous arrangement. She tells me about the "two Google guys" who attended BeaverCon. They stated their intention to fund a competitive beaver placement market, starting at $2 million a year. Neither of us is clear what this is, but $2 million is a lot of money, particularly when beavers work for free. Their stated goal is to reduce the impact of water usage from their server farms. The scalability of this rodent solution attracted them to the current research at Utah State.

Alves explains that she'll be researching all relocation efforts that have taken place in the United States, Canada, and Great Britain. Her goal is to provide a definitive accounting of what works and what doesn't. This highly utilitarian goal fits well with Alves's purposeful nature. I feel that only something this applicable would tempt her to stray from her beaver wrangling work. But the break is just temporary, she tells me. She is taking a leave from her job, and the door is open for her return. It's likely she'll be back the following summer to check on some things.

Adequate space is the primary requirement of any beaver-focused restoration effort, and I ask Alves if the tribe is purchasing any land. She says there is interest, and people are looking into it. However, it's not clear that casino revenues can be significantly diverted into purchasing beaver corridors. Perhaps this is a role for Google, we decide. It hinges on whether the value of the ecosystem services exceeds the cost of the land. Given the right conditions, beavers will hold up their end of the bargain.

CHAPTER 6

WISCONSIN TROUT

Restoring Driftless Area Streams and Mitigating for
Effects of Climate Change

BEST WESTERN, RIVER FALLS, WISCONSIN: TROUT UNLIMITED CHAPTER MEETING

If the tired refrains from "Margaritaville" emanating from the two-man band at the Best Western Campus Inn bar don't drive me to drink, the information to be presented tonight certainly will. I have come to River Falls, Wisconsin, a college town of 15,000 that sits just thirteen miles from the Mississippi River, to attend the chapter meeting of the local Trout Unlimited (TU) group. I was drawn to this western edge of the state after reading a scientific article on Pine Creek written by Kent Johnson, a long-time TU volunteer. Johnson invited me to attend this presentation, meet chapter members, see some trout streams, and watch the zoom presentation on expected impacts to Wisconsin trout based on the latest climate change predictions.

After winding through the hotel lobby, I quickly locate Johnson, and he introduces me to his fellow chapter members. Sixteen people from the chapter have gathered to watch the presentation together. I had assumed that trout anglers were solitary in nature, but this group seems to want to connect with their own kind. Local craft beer flows, and a raffle is conducted for a rod, some trout flies, and other gifts. Once chapter business has been addressed, the presentation begins. Matt Mitro, a coldwater fisheries research scientist with the Wisconsin Department of Natural Resources (WDNR), was asked by the chapter to give a presentation based on a paper he and a team of WDNR and U.S. Geological Survey scientists published in 2018. I was surprised to hear that Mitro would be giving this presentation because the discussion of climate change was forbidden for eight years under the previous administration. It has been two years since Scott Walker was voted out in an exceedingly close election.

Mitro explains that thirteen general circulation climate models were averaged as inputs to the predictions for Wisconsin trout streams. These climate models are used for predicting both air temperatures and rainfall. This in-depth analysis is summed up by colorful maps of the state: green lines marked streams that could support trout and red lines marked streams unable to support trout. The map today contains extensive threads of green throughout the southwest, northern, and central parts of the state. The projections for 2050 showed the green network of cool streams retreating into a more concentrated area in the southwestern part of the state and toward Lake Superior in the north, with few green lines in the central part of the state. Red and orange lines dominated the state; the color-coded message is clear. Streams will be getting warmer, and yes, we need to stop the actions that contribute to this alarming vision of the future.

As with many climate scenarios, the picture does not provide precision on an individual stream scale. For the attendees in the room, it was difficult to know whether their favorite trout stream would be without trout. On a regional and statewide scale, the picture is much clearer. Their modeling predicts a 68 percent decline in stream reaches that will support brook trout and a 32 percent decline in stream reaches that will support brown trout.[1] This discrepancy stems from the fact that brook trout are slightly more temperature sensitive than brown trout. However, precipitation patterns also factor into the predictions. The western part of the state is expected to see greater rainfall. In general, this can be beneficial to trout because rainfall resupplies aquifers that in turn feed the cool springs that sustain these fisheries. However, strong storms and floods in the winter and early spring can flush out the eggs laid by trout in the fall. A few continuous years with these types of storms can decimate a trout fishery. For the northern and central regions of Wisconsin, an increased incidence of summer drought is predicted. The combination of low summer water levels and increased air temperatures can lead to stream temperatures that exceed the thermal requirements of trout. Additional stresses to the system include major groundwater withdrawals from high-capacity wells that can pump up to 10,000 gallons a day. In some parts of the state, streams and lakes have run dry due to these depleted aquifers.[2]

Despite this alarming projection for the state at large, the Driftless Area is better shielded from climate change than almost anywhere else. Trout streams in the Driftless Area are fed by springs. These springs discharge clear cold water that is ideal for brook trout. However, it is uncertain whether changes in precipitation patterns in combination with increased air temperatures could raise the temperature of these streams in the future. Mitro attempts to simplify the complicated picture, warning that "groundwater is going to be the key factor to keep trout in streams." Knowing that climate pressures will certainly affect Wisconsin streams, the WDNR has created a "brook trout reserve" program identifying streams most resilient to climate change and is conducting any needed restoration work on those stretches. Although this seems like forward thinking, the uncomfortable follow-up question is whether these reserve populations will ever be able to recolonize their former reaches.

The assembled anglers seem to be absorbing the information and the distressing picture it paints. Given the somewhat muted response, I gather that this isn't the first time they've been exposed to these issues, but their questions reveal that they'd like some clarity on this concerning news for anglers. One man accompanied by his wife asks, "Basic question: Is it more advantageous for our trout streams to have dry or wet summers?" Mitro tells him that it is complicated and that there is no simple answer; it depends on how close to a spring one is. Closer might lead to consistent, cold, favorable conditions. Further downstream, water temperatures may rise more rapidly due to shallow conditions that come with dry summers. "Will climate change increase groundwater temperatures over time?" Mitro is asked. He responds that there is consensus that this will happen, but to what extent "we just don't know."

A final question is: "Beaver: good thing or bad thing for trout?" Although this is another question Mitro can't answer definitively, he says he has started a study to answer this question. He has found streams where he can remove a beaver dam and monitor the impacts as well as a stream that he can let beavers recolonize. If the beavers cooperate, he expects to have some data to help answer how changes in stream connectivity and water temperature affect trout populations. The jury is still out on beavers in Wisconsin.

Although alarming, the picture in this part of the state does not fit the typical arc of decline that permeates most environmental narratives. In the Driftless Area of Wisconsin, trout streams were decimated in the early 1900s due to poor farming practices that caused widespread erosion of hilly farmland. Streams became deep chasms, and native trout could not survive in the sediment-choked channels. The first real soil conservation initiative was pioneered in nearby Coon Valley. With help from Aldo Leopold and others, the Soil Conservation Service was able to convert 40,000 acres of a 90,000-acre watershed to contour strip farming and conservation tillage. These efforts resulted in massive improvements in the streams that also increased the productivity of the farms. The predominantly small-scale farmers found that the recommended contour strips of hay running parallel to the slopes were effective in stabilizing the soil. They also provided useful feed for their cattle. Eventually almost 95 percent of the watershed was enrolled in some type of conservation practice. But the conditions had been so compromised that early fish biologists restocked the streams only with brown trout because they thought conditions for brook trout were not likely to return.

Today, sitting in bar at the Best Western Inn in River Falls, the issue of climate change seen through trout takes on a different form depending on your perspective. One can feel proud of past accomplishments and the relative abundance of trout but also feel deeply troubled about the impacts of climate change. The problems of poor farming practices could be addressed with adequate funding and tactful communication. The problems of climate change cannot be solved by a handful of enlightened Norwegian farmers. The solutions require nearly everyone's participation. In our current times, this seems like a tall order.

The meeting winds down, people say their goodbyes, and they head out to their cars. Winter is the most solitary time for anglers, and many will be digesting this somber news for several long months to come. They may distract themselves with tying flies or the more ubiquitous distraction of watching football. Through the removable walls that delineate our conference room, I hear the band run through a particularly bitter version of Foreigner's "Cold as Ice." I pay my bill, settle my plans to meet with Johnson tomorrow afternoon, and head out into an unseasonably warm December night. As I walk away from the bar, the lead singer won't release me from his accusatory, auditory grip, reminding me that "someday you'll pay the price, I know!" Yes, I know, I think to myself, and seconds later his backup signer gives voice to my thought. If the science does not convince the residents of Wisconsin that trout are in trouble, perhaps the subliminal and played-out refrains of classic rock will.

DUNN COUNTY: HEAVY MACHINERY AND
AMBITIOUS RESTORATION GOALS

The GPS coordinates that Nate Anderson provided send me east on county highways that wind through rolling hills interspersed with farmland and smaller patches of forest. In this part of the state, the bluffs at the top of the hills tend to be cultivated, and the hillsides that fall away remain forested. This is a disorienting inversion of my understanding of a landscape. I'm accustomed to places where the lowlands are flat and tamed and the elevated areas remain inaccessible and untouched. Here the valleys continue downward in a sinuous fashion, their terminus not clear. As I enter Dunn County on this December morning, I begin my decent into the headwaters of a creek system that becomes deeper and wider as I drive. I am enveloped in a brownish-gray landscape devoid of any contrast save the bigtooth aspen that paint chalky white stripes on the otherwise drab hillsides. It's a muddled picture, painted from a color palate that expands and contracts with the seasons.

I am heading toward an area where several recently completed and planned stream-restoration projects are tucked into the northernmost regions of the Driftless Area. This area occupies portions of Wisconsin, Minnesota, Iowa, and Illinois but is centered in southwestern Wisconsin. Recent glaciations missed this area, so the land wasn't scrubbed flat by the millions of tons of unconsolidated soils, rocks, and debris (drift) that were pushed over and eventually deposited most everywhere else in the Midwest. By missing this date with glaciation, the area gained much more. Deep valleys, numerous spring-fed streams, and rolling plateaus that seem out of place when compared to the rest of the upper Midwest define this area. Although somewhat of a secret to many in the region, it's no secret to trout anglers.

Anderson is the head of the self-described "trout strike team" in the WDNR for this portion of the northern Driftless Area. Johnson had put me in contact with him saying, "this is the guy you want to talk to." He and his crew of four are tasked with the earthmoving and sculpting of trout streams. Working closely with a fish biologist in the WDNR resource management group, they select streams they believe could be improved by restoration. Anderson suggested we meet at South Gilbert Creek, one of several tributaries to Gilbert Creek that eventually flow into the Red Cedar River and the Chippewa River before emptying into the Mississippi River in Pepin County. He tells me that they are prepping this site and that we can visit other streams nearby that have been restored in recent years.

When I arrive at the site, Anderson and his crew member, Josh, and Kasey Yallaly, a fish biologist for WDNR, greet me. Anderson wears a bright green reflector vest, jeans, and a Department of Natural Resources ball cap. Both Yallaly and Josh are suited up in Carhartt jackets. Yallaly and Josh appear to be in their late twenties or early thirties, and a graying beard suggests that Anderson is moving into middle age. They seem excited to show a newcomer around but also self-conscious about the overall appearance of the land. "The is the ugliest time to visit the project," Anderson says.

A large area between the stream and the road has been scraped clean of vegetation, and the bare, rich bottomland soil is thawing on this warmish December morning. Josh has cleared this vegetation, mostly unwanted box elders and willows that must be removed to allow for the placement of rock and eventual grading work. "We rock heavy so we don't have to come back," Anderson had explained earlier when he prepped me for the visit: 1.3 tons of shot rock, a combination of boulders and cobbles of mixed size, will be delivered for each linear foot of stream restored. They construct in sections ranging from 1,500 feet to 6,000 feet. When they stage rock for a project, they store anywhere from two to eight thousand tons of rock. Anderson tells me this is necessary because of the increase in larger storms that climate change has brought to this part of the country.

Rock makes up the largest cost item in the overall project and is critical for ensuring the stability of the streams. Fortunately, the proximity of several quarries keeps delivery costs reasonable. To avoid the ire of these suppliers, who would likely stop supplying them if they were to lodge a dump truck in a soft, muddy floodplain, they must have the rock delivered while the ground is frozen. Although Anderson and his crew can't control the weather, they can prepare the site and be ready for the relatively narrow window for work in the channel that begins May 1 and runs through October 15. When Anderson's team goes into action early in May, the trout eggs will have hatched and the young of the year will be mobile enough to find refuge upstream or downstream.

This period looms large for Anderson and his crew. Floods and equipment breakdowns can limit the amount of work they will finish. His goal of completing two miles each season requires everything to move smoothly. A typical summer might include two or three separate project sites that require separate staging and negotiations with different landowners. They have been working in these tributaries of Gilbert Creek for several seasons. Since 2002, more than four and a half miles of stream have been restored in this watershed. Yallaly, the fisheries biologist, mentions that this area makes up one of their brook trout reserve areas, an initiative backed by both the WDNR and Trout Unlimited. By concentrating their efforts in these streams with good "thermals," they hope brook trout can be sustained here, at least in this headwater area, for some time.

We stand on the bank of the unrestored stream. The morning sunlight rising above the valley walls hits our eyes and illuminates the bare earth. We either shield our eyes with our hands or look downward. The denuded ground provides an unimpeded view of the stream channel and its four-foot-high banks that, in places, sluff dejectedly into the stream. Still lying in the shadows, the stream looks beat up—wide sections of the channel that once carried the active stream now sit dry and purposeless. Josh, who also operates a massive dozer, points out a couple of reds, subtle depressions in the gravel riffles that have been scooped out by trout when laying their eggs.

I ask Anderson and Yallaly what a good outcome would be for this project. Anderson breaks it down. "Taking a stream with 500 trout per mile and poor recruitment to one with 3,000 trout per mile. That's the gold standard." Yallaly, a one-time veterinary science student who pivoted to fisheries, nods in agreement. She has made these counts in other restored streams, having found that shocking fish is preferable to cutting open mammals. Numbers are not everything though; they want to specifically increase the brook trout population six-fold with this work. One of the primary obstacles are the previously introduced brown trout, which are more aggressive and can push brook trout out of stream segments. Browns are also prized by anglers because they grow much larger and are enchantingly elusive, but brook trout are what belong here. Creating the ideal habitat for this native species is their collective challenge.

As we walk along the stream, Anderson tries to show me what he will be doing during the summer. He frames the channel in front of us with his hands and begins to sculpt the air with his hands. There are no plans to reference, this is "field engineering" he tells me. Anderson explains that this sort of work is best learned by doing and that after a while you understand what the stream needs in a particular area. He learned this work in exactly this way, initially being told to "jump in that machine and get to it" by John Saurs, his retired mentor. "If it works out, then that's my doing; if there's a problem, then that's Josh's fault," he jokes.

A typical project requires five pieces of heavy machinery. Two excavators, a dozer, a skid steer, and a track truck make up the heavy munitions of Anderson's platoon. First, they use an excavator to grade the new channel to a 4:1 slope that will enable higher storms to escape to the floodplain. Then they deploy a dozer to smooth the slopes to give them a "nice, natural, smooth look." Next they load the track truck with rock, deliver it to the stream, and use the other excavator to place it. They cover the rock with soil using the dozer and do final grading with the skid steer. Although he explains the process with military precision, I understand that this operation is no blitzkrieg. Given the sheer tonnage of rock and the length of the stream, this process takes some time.

As we talk, it becomes clear that it's not all heavy machines, horsepower, and rock. Looking over the barren floodplain, Anderson points toward a mangled tree trunk with its roots still attached. Although it got a bit roughed up during Josh's work, it is perfect for anchoring an outside bend and will be great cover for fish. With some intent hand-waving, Anderson indicates that the trunk will be buried into the bank with its mass of roots extending into a pool. Anderson also points toward the bend in the stream in front of us. "Right here I'll grade back this bank, shore it up with stone to prevent movement, and turn this area into a cobble riffle." In this section of the river with a gentle slope, no grade control structure will be needed, but in others, he might install several carefully placed boulders to concentrate flow and create a pool. He mentions that they seed the soil with a prairie mix, although he is probably not the best person to talk to about that as he's not a plant guy. As he describes how he intends to create a channel that will withstand storms an altered climate will almost certainly bring, the work seems straightforward, almost preordained. But knowing what can go wrong, I realize that this work requires a certain fearlessness.

Anderson mentions that they've altered their approach in the past couple of years because of Yallaly's influence. They now create more in stream islands and use more root wads to support brook trout. They've stopped using the strong wooden crib structures known as "LUNKERS" (Little Underwater Neighborhood Keepers Encompassing Rheotactic Salmonids) that form a sort of underwater hotel for fish by providing both cover and depth in the banks. These structures were used in Pine Creek, Anderson tells me, a project he worked on with his mentor in 2007. Saurs taught him pretty much everything, but this is one lesson he is unlearning. These structures turned out to be ideal habitat for brown trout.

They spend between $20 and $50 a linear foot on these projects. I'm stunned by this level of productivity and cost-effectiveness. As I think back on projects in the District of Columbia that cost more than $300 a linear foot, I'm not certain what to say. Even the mention of it will confirm every preconception of bureaucratic waste.

Efficiencies arise when you can eliminate hours of time reviewing and editing plans that, during construction, act more like guides than precise blueprints. The bare soil in front of us would, in most construction projects, require costly and time-consuming silt fencing. "There was a time when we had to put in that silt fence *and* take it down every day," Anderson tells me. "Because we were hauling in rock, we had to install silt fence around our rock piles. So we were protecting the rock piles from washing into the stream, but it was fine for us to put the rock in the stream ourselves." These sorts of illogical and labor-intensive requirements can be a source of annoyance, particularly if you are intent on completing two miles of stream each summer.

I think about this level of productivity in comparison with some of my projects. I remember, in the gauzy way one remembers endless bureaucratic hurdles, that a two-mile project I managed took eight years to complete. I share this embarrassing fact with Anderson and Yallaly. "That's why we work here," Anderson says, nodding in understanding.

I ask Anderson about the partnership with Trout Unlimited (TU). He says that they've been a huge help in removing unwanted box elders and willows along the stream prior to restoration. Despite being native to North America, box elders have been deemed invasive and a nuisance when found near trout streams. On the TU website, I had seen pictures of members engaged in a variety of workdays, ranging from tire cleanups to the removal and burning of unwanted trees and brush. The muscle TU has provided helps reduce the overall cost of the projects. They've also contributed significant funds toward projects over the years, either through their membership or through grant writing. Anderson seems comfortable with these partners weighing in on various aspects of the project: TU providing volunteers and monitoring, Yallaly recommending certain channel features, and the landowners closely watching over it all. It is a delicate balancing act that he seems to have fully under control.

We talk of other streams I should visit, and Yallaly excitedly suggests nearby Cadey Creek, a restoration project completed last fall that adds to this growing brook trout reserve. Anderson agrees and says it's just a short drive south from here. I mention that Kent Johnson is going to take me down to Pine Creek tomorrow. Yallaly looks down and pauses, as if I'd brought up the wrong topic at the dinner table. She says that the Pine Creek stream has probably already flipped from brook trout to brown trout. Before this trip I was generally in favor of all trout. Now it seems I must pick a side.

TROUT OR BEAVER: ONE HAS TO GO

Downstream less than a mile we're staring into a pool in the main branch of Gilbert Creek, a previously restored section. The stream runs narrow here, and its banks are blanketed with tall grasses wilted brown at this time of year. We stare down into the clear waters, but the low early morning sunlight and turbulence in the stream prevents us from seeing the bottom. I look to Yallaly as if she can magically tell me what type of fish lurk below.

With the expanse of grasses in the floodplain interspersed with small patches of forest, this entire bottomland evokes a sense of immutability. It's a notion quickly debunked when Anderson tells me this area used to be inundated with beaver ponds and has been restored in the past five years. In diametric opposition to the efforts in Washington, restoration here involved removing the beaver dams and creating greater connectivity for trout. The bright morning sunlight reflects off the narrow, single-strand stream as it winds through the open bottomlands. It aligns with my sense of what a trout stream should look like, and Yallaly tells me that they've counted 3,000 trout per mile in this stretch (figure 6.1).

As we head out, a truck pulls into the parking lot. It's a U.S. Department of Agriculture Animal and Plant Health Inspection Service (APHIS) staff member who Anderson and Yallaly recognize. Yallaly says that he is on contract with WDNR for beaver removal. Apparently, the beavers still lay claim to this stream. He explains that he thought he had taken care of the problem a week ago by removing two pairs of beavers from the lodge down at the road. Based on some recent activity, he now believes there are more. Yallaly says she thought the water level looked high. He assures them both that he will get the remaining beavers, but he needs to wait for the person with the private trap to remove theirs. Everyone nods. These Wisconsin beavers will not receive the same treatment as their Washington kin. These are kill traps.

Yallaly needs to head back to her home office, and Anderson wants to show me another restored section, so we head back to the truck. We cross the bridge the APHIS contractor was talking about: a pile of three-foot-long sticks has been thrown up on the edge of the river. Someone either excavated the lodge or the beavers put up a fight. At the bridge, we look upstream at an unrestored section. Anderson points out the multiple branches of box elder jutting out from the stream banks. Although it would be difficult to navigate a canoe through this

6.1 Nate Anderson, habitat specialist, and Kasey Yallaly, fisheries biologist, look out over a restored section of Gilbert Creek in Dunn County, Wisconsin.

section, the channel seems stable. There are no signs of fallen trees causing undue erosion, but given the oblique angles of the trees extending across the channel, it is not difficult to imagine them falling into the stream—that is if the beavers don't get them first.

CADEY CREEK: A STREAM RESCULPTED

Anderson pulls the truck off the country road, and we get out. We are at Cadey Creek, the next drainage south that eventually drains into the Chippewa River before emptying into the Mississippi. He points toward a stand of dense hardwoods where the stream is cutting into a steep hillside. He shrugs and says they couldn't obtain an easement on this side. He tells me that the owner uses the area as his private hunting area. "I get it," Anderson says, revealing that he has more than enough work on his plate. He explains that they "work where they can." The easements that WDNR requires before undertaking these projects is a policy created in the last five to seven years. "We figured if we are spending lots of money on these projects, people need to fish them forever and not just benefit the landowner in the future," he explains. The easements require that the landowner allow fishing access, keep livestock out of the stream, and allow any maintenance activities that the WDNR deems necessary. Anglers generally stick to the river and the typical maintenance is seasonal mowing to keep unwanted woody vegetation at bay, so it seems like a minor inconvenience for the landowner. In exchange for offering this access, WDNR pays from $2,000 to $4,000 an acre based on a formula that weights issues such as proximity to public lands or major population centers, the presence of structures, and whether both sides of the stream are included. A farmer who owns a soggy forty-acre parcel with a stream running through it might reap a one-time payment of $30,000. Anderson says the completion of a project frequently brings calls from neighboring landowners now interested in obtaining an easement for their section of stream.

Obtaining easements on stream-restoration projects is not uncommon in other states, but is not always required. In many areas, restoration is prioritized on public lands where ownership is in the hands of the government. WDNR's payments have ensured that projects are generally welcomed by landowners. The easements have made anglers happy and provided easy access to streams otherwise inaccessible. The standard easement width is 66.5 feet, but it varies depending on topography, roads, and other property issues. Anderson mentioned that a native prairie mix is used during seeding but acknowledged that to properly establish this habitat type regular burning or mowing in combination with invasive control is necessary. Due to limited budgets and the lack of staff dedicated to this sort of work, the easement corridors can become dominated by invasive plants such as reed canary grass. In my earlier discussion with Johnson, he said this is one area where multiple benefits are being missed, including creation of habitat for pollinators. Johnson also acknowledged that most TU members are more interested in helping with the next stream channel reconfiguration than providing ongoing weeding and time-consuming burning efforts.

After this explanation, we cross the road and look at the stream valley below. This project was completed late in September, which is later than they usually work, and the vegetation hasn't come in as strong as he had hoped. The banks have been gently graded back into the surrounding fields. The grass adjacent to the stream is surprisingly green even though the surrounding landscape has turned brown. With its monotone color and lack of trees, the site reminds me of a tidily maintained golf course. The stream runs pellucid through a channel that is armored with boulders and cobble. Root wads jut out of the water in some locations and remain submerged in deeper pools in others. We head down the hill toward the stream for closer inspection.

As a relative newcomer to the state, the streams in this region are a welcome surprise to me. Anderson tells me he was born in nearby Menomonee and has lived here most of his life. Aside from four years in college in nearby River Falls where he picked up his degree in resource management and some years living outside the Twin Cities, this rural area is where he works, plays, and raises his son. Although small town life can become claustrophobic for some, Anderson isn't itching to go anywhere. Being in charge of reshaping many of the local rivers and streams negates the sense that nothing much changes in small town America.

We stop at a small island that lies in the channel. Nate tells me that these are the habitat features Yallaly likes to see. The stream widens and separates around a small island that, in this recently completed state, is covered with patches of grass and woody debris (figure 6.2). The spot reminds me of any number of

6.2 Nate Anderson shows off a recently restored section of Cadey Creek.

islands I have seen on canoe trips and river walks over the years. It strikes me that it is easy not to notice what appears natural. What adheres to your concept of a landscape possesses a sense of appropriateness. Although most restorationists would rather talk about numerical targets than general appearance, the fact is that this work also serves a community of anglers who have very specific and at times fetishistic notions of streams. He points out a row of boulders that forms one side of the island. "Kasey asks me to make these look messy," he says. "I'm trying to shake that habit of making clean lines." Looking at this river separating at the point of the island and gracefully reconnecting twenty feet downstream, I realize that operating heavy machinery to create a natural-looking island might feel like an unnatural process.

As we walk on, Anderson points out a small area of erosion at the base of a swale he created to help drain a farmer's field. I probably would not have noticed it if he had not brought it up. For someone who joked about attributing the mistakes to his staff, this morning Anderson has pointed out three areas where he felt he didn't get it just right. I remember his earlier description of training in the dirt crew where "we learn from mistakes," and I suspect he is a harsher judge of his work than he lets on. In a field where fingers are quickly pointed and blame freely attributed, I find this tolerance for imperfection and improvement refreshing. When your job site is defined by water that can create unpredictable events, it is a healthy perspective to take. The results in front of us indicate that the iterative process works.

After we look at some stone structures that concentrate flow and create pool and run habitat, Anderson asks if I want to continue. "It just kind of keeps going on," he says, as if he's showing off additional rows of his cornfield rather than a newly created stream channel that may support up to 3,000 trout each mile in the coming years. I ask whether he ever fishes the streams he restores, and he says he prefers lake fishing. "When you spend forty hours out here, you kind of want to go somewhere else on the weekends," he says. I find his honesty disarming. Despite the mythic allure that trout possess, for some of us it's enough to know that they're there. Not all stream restorationists need to spend their weekends back in the stream reading hatches. Perhaps worried that he sounded the wrong note, he adds, "We do occasionally see anglers in sections upstream of our projects. It's nice to see people enjoying the streams."

For all his talk about making mistakes, Anderson is intent on getting things right. There is no small amount of courage involved in having your work visible to all during the messy construction period and open to perpetual evaluation after your work is done. To achieve the results that Anderson wants, he is relying on rainfall to sustain the buffer plantings but that does not come so intensely that it erodes the soil not yet secured by roots. He is hoping that brown trout won't colonize the newly created habitat and that brook trout will find his island acceptably untidy. More than anything, he is hoping that these interventions will help sustain a trout species that is swimming upstream against future climate impacts.

MORE THAN MEETS THE EYE: KINNICKINNIC RIVER AND UBER-VOLUNTEER KENT JOHNSON

Kent Johnson's appearance doesn't stand out in this part of the state. Where his beloved brook trout possess olive-green vermiculations that resemble refracted sunlight passing through the rippled surface of a trout stream, Johnson has jeans, a blue sweatshirt, and a baseball cap. Add in his soft-spoken demeanor and he blends into this midwestern landscape. He's athletic for a retired man of his age, and the lines of his face are just starting to reveal that he may be thinking of slowing down. But this afternoon he's eager to show me several of his monitoring sites in River Falls.

We're standing next to Johnson's 4Runner in a dirt parking lot next to the Kinnickinnic River just off the main drag that heads into downtown River Falls. Last night's dusting of snow is melting away on the railings of the bridges we just crossed to get here. In one direction, the QuikTrip and numerous fast-food restaurants line the road into town. From the other direction, the noise from Highway 35 drones softly in the background. This ubiquitous, sprawling, car-centered development is what irreversibly degrades our streams. An often-cited study from the Center for Watershed Protection, a national leader in watershed science, claims that at 10 percent impervious coverage stream health declines and is nearly impossible to reverse. Stormwater inputs doom the pollution sensitive macroinvertebrates and that initiates the decline in fish populations. Looking at the stream and the surrounding environment, I have a leaden sense that I've seen this story before.

As we stand on the banks and survey the stream, Johnson begins to explain what he has been doing. He tells me that this stretch of river is a class-1 trout stream and that people can successfully fish this section for nice browns. His bit about fishing in the city proper doesn't quite sink in as I look for signs of tires and litter. Other than a stray wrapper in the parking lot, I don't see any debris. Upstream the banks are forested and stable and the water runs clear. I look toward him and ask, "Did you say people fish here?"

Class-1 trout streams are the healthiest of all trout streams. This designation indicates trout are restocking themselves and the stream provides the required habitat and macroinvertebrates to sustain them. Most trout streams are found in the uppermost reaches of stream systems, the proper conditions existing only in the dendritic tips of river systems high in the forested hills. Answering my question, Johnson confirms that people do fish here even though the popularity of river kayaking has increased overall traffic on the river. Anglers tend to come out early to avoid the crowds. I ask if any other class-1 streams run through cities in Wisconsin. He says that he is not aware of any.

Perhaps recognizing my slow uptake, Johnson deliberately describes his network of monitoring locations. He explains that the site in front of us is his upstream temperature monitoring station, which he has been checking for almost thirty years now. This allows him to compare stream temperatures at this location, upstream of stormwater inputs, to readings lower downstream. Higher temperatures downstream, particularly in the summer, point to the effects of stormwater runoff. In addition to the perilous effects of increased temperature on trout, stormwater runoff carries numerous other pollutants such as sediment, metals, and nutrients that affect stream health.

He and another TU volunteer started with four locations in 1991. In 1992, they began adding locations on two nearby tributaries, expanding their network of sensors. The watershed was their patient, and they took great pains to poke and prod in each reach. These initial stations cost about $750 each, which included intake pipe, pumping devices, and data recording devices. In 2000, Johnson began using a new technology, Onset Tidbit v2 loggers, quarter-sized temperature loggers that could be attached to any hard surface with a cable tie. These devices could capture months of data and cost only $149. Given that TU was footing the bill for this work, the cost savings were significant. Eventually, Johnson and his partner set up nine monitoring stations. Although they only captured temperature data, when correlated with local stream stage gauges, the changes in temperature acted as a proxy for measuring stormwater effects on a reach scale. This geographically specific data record provided evidence of problems from runoff, an ongoing challenge with widespread effects such as stormwater. This specificity would pay dividends.

Although direct measurement of these pollutants can be made, I know from firsthand experience that it is not easy. Collecting samples is challenging, and lab processing costs are steep. One must either rush to a stream at the first sign of a storm or set up fussy automated collection systems that frequently fail and require ongoing maintenance. This sort of direct measurement is typically undertaken only for well-funded research projects or in cities with well-funded stormwater programs.

With a shoestring budget, Johnson is addressing an important issue often raised in the study of streams. During panel discussions at restoration conferences, everyone sagely nods and agrees that more monitoring is necessary. But few seem able to carry it out. Monitoring the trajectory of restored streams and suitable control sites for the duration necessary to capture these long-term trends requires expert staff and long-term funding. Five years of monitoring might point to a new impact, but ten is required to understand if it is a one-time impact or a trend. For impacts that roll out over decades, such as development in watersheds or climate change, decades of data may be needed.

Many academic labs place a premium on assessing the fate and transport of pollutants that are of particular concern. New technologies may become available for new detection or processing techniques that can reveal new trends. These types of highly advanced technologies are catnip to academics; they may reveal new relationships, and the results are likely to be published in academic journals. But the cost and the generally short nature of grants that fund them render them of questionable use in determining what is happening in restored streams. Ultimately, the time needed to discern how streams react to these major disturbances generally stretches out long past the grant periods, the availability of post docs, or tenure considerations of professors.

Streams undergoing restoration projects see intense disturbances and then periods of naturalization influenced by many abiotic factors. Each stream project deals with unique factors and unique activities within the watershed, so only assessment of a great number of streams over a long period of time can identify actual trends. The standard requirement for monitoring, if it is even required, is five years of postconstruction monitoring. This clearly doesn't allow for a complete understanding of how these evolving systems are responding to restoration efforts.

Johnson walks over to his temperature sensor, inserts a flashlight-sized data recorder into it, and tells me, "That's it." I stare and try to process the implications of what he's showing me. Although Johnson could have designed a more complicated monitoring approach, I realize that his long-term, streamlined effort is, in this situation, more powerful. Johnson and his fellow volunteers have built this network using this cost-effective surrogate for temperature readings. It's significant that his efforts are entirely voluntary and funded by an active and relatively small TU chapter that receives no government funding. Although he cautions that direct measurement would be better, by correlating these long-term temperature readings with long-term weather data and stream gauge data, he can understand the runoff from a particular drainage area. Using estimates of pollutant runoff for a representative land use type, he can estimate pollutant contributions. It's not dissimilar from making a case with overwhelming circumstantial evidence when direct eyewitness testimony is unavailable. Whether or not it sways a jury, it paints a very clear picture of what is happening.

Sometimes the jury in these matters can be very small. He tells me that he started deploying these units in the late 1980s after initial discussions with the city engineer of River Falls. The city engineer had come from the larger city of Rochester, Minnesota, and was familiar with new stormwater requirements associated with "phase 1" permits. These permits from the state and overseen by the EPA are the primary method of regulating the entirety of urban stormwater pollution. In the 1980s, larger cities were required to develop plans to attempt to control the multiple sources of pollution. Approaches ranged from training city staff to retrofitting specific projects that addressed pollution "hot spots." One important component in many of these permits was the development of new stormwater ordinances. These ordinances placed requirements on new developments (older developments were usually grandfathered in) for how much stormwater they would be required to treat and detain. Numerous practices have been developed to do this, ranging from stormwater ponds to green roofs. But each requires a developer to set aside funds and, in most cases, space to manage this water. The ordinances typically apply to sites larger than half an acre and are often described by the size of storm that must be detained. In Milwaukee, for example, a recently adopted ordinance requires that sites adding more than 5,000 square feet of new impervious area must capture the first half-inch of rainfall on site.

Johnson tells me that some readings taken over the course of a summer provided evidence that stream temperatures rose significantly as the river passed through the city. This was compelling enough proof that stormwater was having thermal impacts on this resource and that other pollutants were likely being added as well. In addition to temperature data, trout anglers reported signs of sediment discharge from new developments on the outskirts of the city. Johnson says that the city engineer took a good look at his data. After roughly a decade of these sorts of discussions—a period of "mutual trust building," as Johnson diplomatically puts it—the city adopted a new stormwater ordinance in 2002. The ordinance called for the detention and treatment of the first 1.5 inches of rainfall: "Due to its proximity along the spring fed Kinnickinnic River, the City of River Falls has chosen to adopt an even higher threshold in its Storm Water Ordinance."[3] By one yardstick, this is three times more protective than regulations touted in many larger cities. A decade's worth of temperature data from Johnson's $150 units had swayed the jury.

Standing next to the stream and imagining myself fishing here next summer, I ask Johnson again about the stormwater ordinance. I had heard of requirements this strict only in pristine headwaters or in special zones with strong development interest where a city could demand such measures. This very strong ordinance was passed in sleepy River Falls, not known as a hot spot of development. This stretch of river running behind the gas station took on a renewed sense of vigor that challenged my assumptions. Not only was this stream producing and sustaining trout, it appeared to be well-protected from future development. One could imagine a nonextractive economy both supported by and protective of this fantastic natural resource. This dirt parking lot along the Kinnickinnic River behind the strip businesses on Highway 35 had become something strangely special.

BEAVERS . . . AGAIN?

The floodplain of this small creek is covered in chewed off stumps of poplar trees. Johnson and I look at the small creek, dammed up by a beaver, and he shrugs, apparently not concerned. As with my earlier run-in with this elusive creature, I'm not sure what to say. I want to focus on water quality and trout, but the beaver's influence stares us in the face. This tributary to the Kinnickinnic is called Sumner Creek. It is one of the coldest tributaries and supports a diverse macroinvertebrate population and brown trout reproduction. This is a critical resource, Johnson tells me.

The headwaters of this creek are not the forested and remote reaches of some other creeks. The source of this pristine creek is a former farm that is bisected by the highway. In the middle sits a major development called Sterling Ponds. Construction on this 158-acre housing development with 482 platted lots began just after adoption of the new regulations.[4] What happens in this creek that seems to be visited only by Johnson and a handful of beavers will test whether the new stormwater regulations of River Falls can truly protect the vibrant fishery in the Kinnickinnic. I want to come face to face with this potential threat. In a field that is awash in concepts and trends, this development represents a very direct experiment; namely, can we foster economic development and still protect our water resources? It's a scenario that is replicated throughout the country but rarely studied directly. We ignore the beavers and head upstream.

STORMWATER IN SUMNER CREEK: CAN DEVELOPMENT AND TROUT COEXIST?

The place still has the feel of a new development, as if it has just popped up from the surrounding farmland overnight like mushrooms after a good spring rain. The curved roads that wind through it are wide, and the homes are large. Recently planted trees in neat lawns suggest that the homeowners intend to stay for a while. A new home is being constructed at the end of the road close to where we park.

Johnson takes me behind the row of houses that abuts a large wetland dominated by cattails. In the back of the yards, a level grassy field at least an acre in size sits below the homes and about two or three feet above the elevation of the wetland. Johnson explains that this is one of the infiltration basins for the development. At the far end of the field, a pipe emerges from the base of a berm. As I follow Johnson, I notice that the berm forms the base of a stormwater pond. We stop at plywood box held up by a couple of 4×4s reminiscent of a crude wild west gallows structure if not for the rain gauge perched on top. This station, labeled "City of River Falls Water Monitoring," is situated right next to a large round grate (figure 6.3). Peering down, I see that the grate covers the buried overflow

6.3 Kent Johnson, Trout Unlimited volunteer, at the Sterling Ponds development stormwater monitoring site for Sumner Creek.

pipe that runs from the pond to the wetland. This expanse of cattails is the headwaters of Sumner Creek. Having seen the components and their connections, it's now clear how the developer met the local stormwater ordinance.

Later that day I look over Johnson's 2014 monitoring report and try to determine whether the new stormwater ordinance is indeed protecting the nearby creek. For those unaccustomed to reading monitoring reports, they can be mind-numbingly detailed. In this one, Johnson's thoroughness is evident. Using his $150 temperature units and a sensor at the overflow, he has developed a monitoring approach that allows him to determine exactly how much rainfall is going into the basin. This inexpensive approach provides a reality check for calculations that form the basis for site plans. I focus on one section in which Johnson presents data from the site we visited. The year 2014 was particularly wet and, as could be expected, not every storm was diverted to the infiltration basin. That summer saw three large events greater than 1.5 inches.[5] Those and one other event near 1.5 inches resulted in direct overflow from the pond to the creek. However, considering all rain events over the summer, 70 percent of the total rainfall was directed to the infiltration basin where nearly all pollutants were removed and any thermal impacts were buffered. The fact that this very aggressive stormwater system still allowed 30 percent of the total rainfall to bypass the system speaks to the challenges of managing stormwater when extreme rain events become more common.

I imagine this uncaptured 30 percent. It came during the summer when a warm front dumped buckets of rain on the landscape. The warm rains picked up oil from driveways still hot from the day and sediment that flowed off bare sections of yards. Did these flows, the missed 30 percent, affect the trout populations in Sumner Creek? In last night's presentation, Mitro indicated that trout can withstand temperatures that exceed their thermal tolerance if the period lasted for only a couple of days. Although the stormwater ordinance has been met in this new development, only an infiltration basin the size of the entire development would ensure that every storm is captured. The variability of future storms in an altered climate presents unknowns that engineers struggle to accommodate. As we look over the headwaters of Sumner Creek, I look to Johnson for some clarity. "Is the ordinance going to provide adequate protection for the trout of the Kinnickinnic River?"

"Big picture, yes, the ordinance is doing its job," Johnson says. "With the more intense storms of the future, we might need to look at upping the retention standard, but it shows that in communities such as River Falls where there is adequate land we can capture most of the rainfall and protect cold water resources." Having seen endless developments with stormwater ponds that don't adequately protect the downstream waters, this suburban development, carved out of a former field, takes on a new shape. By meeting the new infiltration standard, Sterling Ponds shines as an unlikely beacon of sustainability.

With the participation of a few volunteers over the years, funding from the local TU chapter, and active engagement of local decisions-makers, the data Johnson is collecting is invaluable. It provides a long-term view of how these trout streams in western Wisconsin are handling climate change and stormwater inputs. It provides insight into how stable these spring-fed streams will be going forward in a climate subjected to higher temperatures and more intense rainfall. As I sit adrift in the sea of complexity that is our climate crisis, Johnson's data, if not providing the solution, is at least shining a light on what is happening. This knowledge, if we pay attention, can direct our efforts in the right direction.

PINE CREEK: THE POWER OF MONITORING
AND VOLUNTEERS

The next morning when we meet at my hotel, there is a dusting of wet snow on the cars in the parking lot. Johnson flashes me a quick smile, and I hop into his 4Runner. Our plan today is to head thirty miles south to Pine Creek, an important WDNR restoration project, a long-term monitoring site of Johnson's, and the project that first caught my attention. This is also the project where Nate Anderson cut his teeth.

Johnson sent some of his monitoring reports to me the previous day, and I had pored over them so I could pepper him with detailed questions. As we head south, I'm most curious about why he continues to monitor streams after having spent thirty-eight years doing it for the Metropolitan Council of Governments for the Twin Cities. I attempt to get at the issue by asking how he likes retirement. It's one of those mundane questions people tend to ask, but I'm not caffeinated enough to come up with anything more original. In his taciturn manner, Johnson says he was ready to retire but also enjoyed his work. Noncommittal.

Conjuring my best Barbara Walters, I try again. "So did you enjoy being out in the field more or managing staff?" Johnson isn't used to being the subject of such hard-hitting questions. He's being a good sport though, and he finally comes up with something. He begins explaining a project as we approach Lake Pepin, but then he pulls over at a scenic overlook to describe the scene outside our window. Lake Pepin, unlike many of the backwaters created by the numerous dams on the Mississippi River, is a natural lake created by an ancient glacial river. It is twenty-two miles long and two miles wide and set among the rolling bluffs found on both the Wisconsin and Minnesota sides. Johnson mentions that the bluffs are nice over in Minnesota, "just not as nice as on the Wisconsin side." Just "Minnesota nice?" I clarify.

The lake, placid today, is inseparable from the massive river that flows through it. Its forty square miles of flat water gave rise to the sport of waterskiing. As we take in the Arcadian scene in front of us, the lake seems to be urging the river to take a break from its well-known patterns of industriousness and commerce. Johnson says that this was a unique sampling site for his job in 1998, when the combination of an extreme drought in the upper Mississippi watershed combined with nutrient pollution from the Twin Cities resulted in algal blooms and severe fish kills in Lake Pepin. "It was a bad scene," he relates. Fingers quickly pointed toward the major wastewater treatment plants just fifty miles upstream that served three million Minnesotans. Resultant fish kills pressured the municipalities to find funds to better monitor phosphorus discharges. Johnson's office was tasked with this work. By 2003, phosphorus limits were ratcheted down to 1 mg/L, a level that has prevented similar impacts on the lake.

I ask him if he tried to carry out his stream monitoring on the Minnesota side of the river while he was working there. He said it was easier to involve himself in Wisconsin so there weren't any complications related to his employer or other people with whom he regularly worked. It's a small tell for someone who has probably been trained by years of government service not to push the wrong levers. Although he wouldn't come out and say it, I detect a quiet, transboundary agenda in his continued deployment of equipment in these driftless streams on the Wisconsin side.

We continue along the Mississippi River Road where the multiple modes of transportation of our country are evident, if somewhat disguised by the weathered appearance of everything. The capricious river lurks only a hundred yards down a steep hillside, and despite its present calm, it is mighty in every sense of the word. Between the road and the river lies the railroad, still active today. The road bobs up and down along the bluffs and skirts wood-sided homes needing a fresh coat of paint and businesses that sit only feet off the road. The buildings anxiously cling to the road, seemingly remembering past floods that saturated their foundations. These visual markers, subtle in this long-manipulated landscape, reveal a landscape defined by rivers. There is a vague feeling of industriousness, combined with a sense that time is moving slowly and that one might best slow down and think things over.

My cell service fades, and we soon come to the road that runs up to Pine Creek. Immediately we are immersed in a narrow valley cut off from the outside world. It's cloudy today, but there is no wind or rain and the temperature is unseasonably mild for a December morning. The valley is perfectly serene. I feel as if I've been dropped into some kind of beige cradle, enveloped by the darker hardwoods covering the sides of the valley and placed down on the senesced grasses and forbs that have been softened by recent frosts. In the parking lot we see only one other car—a mother giving her young kids a chance to get out of the car, walk over to the stream not twenty yards away, and perhaps see some trout for the first time.

Johnson explains that the project was made possible in 2002 when the West Wisconsin Land Trust purchased 220 acres of the valley bottomland that had been grazed as pasture by the former owners. At that time, the stream had incised into the valley, leaving unstable banks and a wide shallow channel. Johnson pulls out a binder and shows me prerestoration pictures of him standing in the channel with the banks extending several feet above his head. Although a population of brook trout already existed in this stream because clear cold springs fed it, the habitat was poor. When planning began in earnest in 2006, it was evident to all that this stream had high restoration potential. With the cattle now out of the stream, the only stressors in the watershed were a few small farms in the upper portions of the watershed.

John Saurs from the WNDR, Anderson's mentor, was the primary field architect of the project. Trout Unlimited was a major player in fund-raising and monitoring. A local rock quarry provided key maintenance support by agreeing to mow the area at prescribed times. Based on past experience, the WDNR construction staff went about regrading the banks to a 3:1 slope, seeding and stabilizing these banks, narrowing the channel, excavating deep pools, and installing the now discontinued LUNKERS structures. I ask Johnson if Saurs was a proponent of Rosgen techniques. He shakes his head and, answering with an ambiguity I am beginning to be able to read, mentions that Rosgen isn't talked of much in this area. When I ask who did the designs, Johnson tactfully says: "It's not that they do it by the seat of their pants, it's just that they're able to make the decisions on site. They've done enough of these to know what they need to do."

Begun in 2007, the project was eventually completed over four summers of work, the details of which were faithfully recorded by Saurs in annual project summaries. When I perused the reports, the dry clinical detail necessitates that one brings some imagination to the reading. I can see a younger Anderson being cut loose on a stream that had been beat up by cattle for decades. When the author mentions the incipient head cuts of the northern spring tributary, I envision that an impromptu meeting was conducted to go over the options for stopping it.

What these reports lack in poetry, they make up in detail. They break out the project in feet restored each year and associated costs. At the end, the five sections are summed up with precise accounting: 11,167 feet (2.11 miles) were restored at a cost of $270,273. The exactness acts as a careful prebutting to any angry taxpayer who might grumble about wasteful government spending. At $24 per linear foot, it is difficult to imagine anyone taking issue with the cost. In many places, it would be easy to find projects that cost ten times this amount on a per foot basis.

We get out of the car, and Johnson leads me down a short path to the banks of the stream. As we look out over the stream, I'm struck by how natural this site appears. Ten years after the project was concluded, there are no signs of past construction work. The stream valley is dense with grasses and forbs, and the banks are stable. In some locations, a few angular boulders reveal where an outside bank was reinforced. We peer into the stream and immediately see trout streaking away. The water is extremely clear, but the pools are deep enough in places that it is difficult to keep visual tabs on the fleeing trout as they dive into their aquamarine world. We walk down a few yards to a LUNKERS structure, and several larger trout scoot under the bank. It's not clear whether they are the same or different fish, but after doing this for ten minutes, I'm astounded by the bounty of fish in the stream. Pine Creek appears to be a field-engineered trout factory, the piscatory output evident at every turn.

Johnson's monitoring paints a picture of a project with numerous successes. The report runs through the accomplishments. The first four objectives were successfully met. The temperature regime was improved through reshaping and revegetation of the channel. There was a 60 percent reduction in stream bank heights, and as a result, the stream is now reconnected to its floodplain. Johnson recorded a 140 percent increase in coarse substrate, critical for spawning habitat, and a 133 percent increase in macrophyte presence, the aquatic plant life that provides cover and basis for overall ecology of the stream. One could conclude that the work of channel shaping was exactly what was intended. It's clear that trout thrive in this habitat.

True to the data, Johnson doesn't skirt the unintended consequences of the project. In charts that indicate the relative abundance of brown trout versus brook trout, he shows how the populations changed over time. Beginning in 2009, after the first sections were completed, brook trout populations exploded. In August of 2011, the nearby *Republican-Eagle* raised the mission-accomplished flag with headline, "Brook Trout Thriving in Restored Pine Creek." By the time construction was completed in 2011, brook trout populations had begun a steady decline and brown trout had gained a foothold. As Kasey Yallaly had revealed previously, by 2014 the populations had essentially flipped, with brown trout now comprising roughly 90 percent of the overall fish abundance and brook trout only 10 percent. In creating a stable stream, the project also created ideal habitat for brown trout.

Johnson's report paints an unbiased picture of the benefits of the project but also alludes to efforts that may have missed the mark. Understanding that brown trout could expand after the project, WDNR conducted shocking and removal of brown trout in 2007 and 2008 before the project began. Apparently not all were removed, although that might not have prevented the eventual dominance of brown trout. It was not anticipated that brown trout might reenter the stream from the Mississippi River, a body of water not preferred by the species. One theory speculates that the browns came over from the Rush River, the next stream north, ducking down into Lake Pepin and then moving into the prime habitat created by the restoration project. Johnson also mentions the presence of interspecific competition, whereby browns push brooks off their nesting sites and can prey on smaller brook fry. The parasitism of gill lice are also mentioned as a source of stress for young brook trout, as if this besieged species needed a case of head lice to add to its worries.

But perhaps the largest oversight was the creation of deep pools and shaded LUNKERS structures that proved to be perfect for brown trout. The structures that WDNR staff had perfected in other streams were executed perfectly in Pine Creek. As Anderson had alluded to, it just wasn't the right type of habitat for brook trout. This points to a common challenge in restoration efforts. Even with talented practitioners, the work eventually is returned to nature, which often fine-tunes the results.

Seeing so many brown trout becomes oddly prosaic, and we head back to the car. I see the trout bag limit sign in the parking lot. "Daily bag limit 3—minimum size 8 inches." Johnson points out the smaller sign just below it, which was tacked on earlier this year by Yallaly. "Brook Trout Restoration Stream: harvest of brown trout is encouraged" it states in a bold font. The sign tells me that Yallaly's not quite ready to throw in the towel.

We head farther up the valley to check out another section of the creek. The road turns to a dirt surface, and we approach a spot where the creek runs right over the road. Even though we've already seen more than a hundred trout in the past hour, Johnson slows his rig as we inch across the creek crossing. Johnson's scientific demeanor morphs into something more spontaneous as he opens the window and peers intently down at the road. He is attempting to spot any trout that may be resting on the clean cobble that sits on the road. It is an odd feeling to be driving through trout habitat, but any fish there quickly dart away.

A bit further down the road, Johnson pulls off in another dirt parking lot, and we walk out to the north spring tributary. Johnson beats a path through the floodplain and stops at a birdhouse he says his father helped make. He and other Trout Unlimited volunteers installed twenty-four of these throughout the project site. They are intended for the eastern bluebird, a species that migrates to the southern United States over winter and prefers this exact open meadow habitat. He opens the flap and pulls out last year's nest made of small sticks and twigs, a clear sign that house wrens used these birdhouses. Last February, Johnson says, an intense cold snap down south killed hundreds of thousands of bluebirds. They are expected to recover, but very few were seen this past summer. As he cleans out the birdhouse, his movements are suddenly tentative, which makes sense when he tells me he has found itinerant mice in these abandoned nests more than once.

I follow Johnson deeper into this place that keeps revealing compelling biological narratives that are somehow incomplete. Eventually we come to what is known as the North Spring tributary, small enough that no map bothers to name it. The stream is too wide to jump across, so I test my Bogs boots in the uncertain depths. Although certain that the creek will overtop my boots, I emerge dry and step onto the dense grasses that line the stream.

Johnson leads on toward the valley wall yards away now. We come to the spring and stop because there is no more stream to follow. Water is pulsing from the base of the hillside, the underwater current visible through the differences in refracted daylight. Verdant mats of watercress float on the surface and trail downstream in the imperceptible current. Against the backdrop of the dun-colored valley, the electric green reminds me of Mitro's maps of projected impacts from climate change. I'm captivated by this place defined and made possible by cool groundwater. Having spent so much time in polluted waters, I stand here watching groundwater emerge from the invisible aquifer that stretches into the hillside. Springs have long drawn people from around the world who seek cures to ailments that defy conventional treatment. As I soak in the visual tonic of this idyllic place, I temporarily run out of questions. I recall how the medical claims of those well-used therapeutic springs were later deemed dubious. But here, at this spring, I sense that I am, if not being cured, then at least receiving some ineffable form of rejuvenation.

I ask how many times a year Johnson comes down to this site. He says twelve times a year, but that doesn't include fishing trips. Sometimes his wife accompanies him. She is a painter who enjoys the site as well. It takes four hours to do the monitoring, about the right amount of time for her to capture the valley. Doing some quick math, I realize that he's been monitoring this site for fifteen years now. I ask if that is enough time to tell the story of the restoration project. "Well, it's likely that the stream has stabilized and many of the measurements won't change significantly." But this admission quickly turns to a discussion of some other interesting things he's keeping his eye on. I have a strong sense that many interesting things will bring Johnson down here for the foreseeable future.

Johnson deployed a temperature sensor at a different spring that feeds this stream further up the valley, and he says the temperatures fluctuate about 1.4°C in a surprising manner. They rise slightly in the winter but drop in the summer. He's careful to state that this would require more research, but he attributes this to a shallow aquifer with a residence time of six months. This delay causes the warmer rains of the summer to eventually discharge at the spring in the winter and, conversely, colder snowmelt to eventually emerge in June and July. Remembering Mitro's take-home point that the springs are the key to sustaining trout through climate change, I wonder if this fluctuation is beneficial or worrying.

He also mentions that he has seen nitrate levels rise over the years in quarterly water samples. Levels have risen from around 2.5 mg/L to 4.5 mg/L. Although the levels are not close to the maximum contaminant load (MCL) of 10 mg/L that would deem it unsafe for drinking, he believes the nitrates give rise to periphyton and filamentous algae that occasionally flourish in the stream during the summer months. There is adequate reason to be concerned about this rise in nitrates. In the county bordering this very stream, more than 20 percent of private wells have nitrate levels that exceed the MCL. In 2012, forty-seven municipal systems in the state had raw water samples that exceeded the MCL. In 1999, only fourteen public water systems contained samples that exceeded this level.[6] Although the effects of climate change loom, the very underground resource that is key to trout as well as human survival is being contaminated right now.

With all of this information, I strain to connect the dots. I've seen many aspects of this work, but I have probably missed others. The relationship between dedicated practitioner and tireless volunteer transmits information in both directions. The impacts from nitrates running off farms combined with time-delayed groundwater flows raises questions of science, but also questions of political will.

The complicated interspecies competition between brown trout and brook trout remains unresolved. Looming over everything is the question of how an altered climate will affect this stream. Every issue that affects the future of trout streams in this state is represented in this one quiet valley. Part of me wants the feel-good ending. I want to remain at the springhead, look at thousands of trout, and simply rejoice in the bounty of what has been created here. It may not have been exactly what was intended, but this valley is teeming with life and points to what can be accomplished with restoration. It also points to how challenging this work is and how factors outside initial project considerations can affect its future.

The full story of the trout in Wisconsin has not been written. Despite impressive efforts by Nate Anderson and his team to create brook trout refuges, these efforts may not be sufficient to mitigate the effects of climate change. In light of such uncertainty, those who continue to try to figure out these relationships and modify restoration approaches are providing essential information. Johnson's ceaseless monitoring reminds us that we would be wise to keep our sensors deployed.

CHAPTER 7

RIVER CANE DREAMS

A Plant That Restores Connections

CRIME SCENE

The black-and-white sign posted by the side of the Andrew Jackson Highway notifies anyone who might stop that this is state-owned property: No trespassing allowed. It's not a place where anyone would stop unless they had car troubles or had some type of nefarious purpose. Should you come here with the latter intent, the sign would offer no resistance. As if attempting to stay out of the way, the sign stands to the side of a wide, grassy access road, announcing its stern warnings while keeping the path clear. As Adam Griffith, the current linchpin of river cane restoration in western North Carolina, and I walk down the road into the scrubby woods, he describes his last trip to this site when he discovered that several large walnut trees had been pilfered.

When he came across the scene, he predicted that some sort of incriminating evidence would have been left that could be used to track the perpetrator. Griffith said the evidence was even more obvious—the illegal harvester had emptied out the cab and left an identifying receipt from the sawmill where he had sold his last purloined load. "I was more upset that he brought in a bobcat and drove over all of this cane," Griffith relates. He reported the incident to the North Carolina Department of Transportation and intended to report it to the police. But after Googling the name of the offender and finding a few prior meth-related convictions, he decided against doing that. That same crime committed in the Olympic National Park would lead to a thirty-month prison sentence.[1] The only result here was this stamped metal sign clinging to signpost. "That's what we're dealing with here," he says.

Camp Creek wanders through this unwatched and ignored parcel before flowing into the Tuckasegee River and eventually into the damned-up waters of

Fontana Lake. We walk through a patch of river cane that lies close to the stream. Cane stalks almost fifteen feet tall extend out from the center along straight tracks, like rays emanating from the sun. On closer observation, it becomes clear that the source in this case are rhizomes hiding from the light a few inches under the ground. The cane is unfazed by the criminal activities in the area. Perhaps thankful for the now thinner forest canopy, it pushes outward into this exposed and unclaimed land.

The cane beckons us to explore. Uniform and relatively free from any understory plants, the stand gives the impression of safety. The fringes are less dense, but it soon becomes difficult to maneuver. The occasional greenbrier (Smilax) vine catches my legs with its thorns, and I'm forced to backtrack. It is humbling to feel trapped by what is, essentially, a patch of grass. Several historical accounts describe the impenetrable river canebrakes. In Teddy Roosevelt's 1908 account of bear hunting in the Louisiana canebrakes, he described them as "well-nigh impenetrable to man and horse." He went on to describe them in more detail:

> The canebrakes stretch along the slight rises of ground, often extending for miles, forming one of the most striking and interesting features of the country. They choke out other growths, the feathery, graceful canes standing in ranks, tall, slender, serried, each but a few inches from his brother, and springing to a height of 15 or 20 feet. . . . It is impossible to see through them for more than 15 or 20 paces, and often for not half that distance.[2]

River cane influenced the outcome of a critical revolutionary war battle. At the Cowpens National Battlefield in South Carolina, dense stands of river cane pushed British soldiers into a narrow strip of uplands where they were routed by American Brigadier General Daniel Morgan. Currently standing in this smaller patch off a highway, I can imagine how these patches might appeal as a refuge when under musket fire, but once immersed, forces would face severely complicated movement and poor communication. With no muskets pointed at me, I'm content that my movement is limited. Being surrounded by hundreds of identical culms, this wall of biomass forces me to slow down. I'm compelled to engage with my immediate environment.

Griffith points out a black and red insect that reminds me of a large pillbug or a short centipede. Its colors jump out in the otherwise monotone shade of this cane stand. "Pick it up," Griffith implores. I'm wary but trust he wouldn't subject me to any type of injury. "Roll it around a bit," he encourages. Cautiously, I do so and then, as he instructs, smell my hand. A familiar and sweet smell of cherry cola hits me, strangely artificial in nature. It's the "Cheerwine bug," Griffith tells me. Even the insects seem to follow some unwritten southern gothic script.

These derelict and unconsidered fringes are where river cane is allowed to grow these days. As we drive down Andrew Jackson Highway, which cuts through historical Cherokee lands, Griffith points out the river cane growing along the river. Growing in thin bands between the highway and the river, its pale green leaves remain throughout the winter and might catch the attention of a passenger looking out the window. Griffith indicates where the river cane transitions to the smaller, and now brown, hill cane as the land rises away from the river.

Not more than a few minutes later, Griffith motions toward another stand along the highway (figure 7.1). For an endangered plant, it seems to pop up

7.1 Adam Griffith points out a vigorous stand of river cane (*Arundinaria gigantea*) just off the Andrew Jackson Highway in western North Carolina.

regularly. Where the highway cuts into the base of a hillside, a small patch of cane hangs onto the 45-degree slope ten feet up the hill. Griffith states that a road construction crew must have inadvertently placed some soil from a cane patch on this hillside during construction. "Steepest location I've ever seen cane grow," he says. In clear view of motorists, I wonder if the plant has selected this location for its apparent safety. Not following the rules laid out in ecological textbooks, river cane follows some southern creed of surviving on the edges of a fragmented landscape.

RIVER CANE MIDDLEMAN

This plant is willing to pick up and move, and appropriately, my guide is also a transplant. Griffith describes himself as a summer camp kid who has stayed. Originally from Pennsylvania, he migrated south to the Smokies, drawn to the dreamlike mountain valleys blanketed by ethereal clouds that give the area its most fitting name. An avid kayaker and road and mountain bike enthusiast, he once rode the entire stretch from his home to his former camp in western North Carolina. Now in his midforties, he has been working with this plant since his graduate days in 2004, and he retains the excitement of a kid anxiously awaiting a full slate of camp activities. Each day brings a new person into this growing cult of cane, and Griffith has graduated to more of a counselor role: part enthusiast, part middleman. He orchestrates numerous interactions between native artisans and a motley group of landowners who are reevaluating their relationship with this plant.

We meet in his office in the converted trailer off the main drag in Cherokee, North Carolina, that houses the Cooperative Extension Office of the Eastern Band of Cherokee Indians (ECBI). This simple structure is the hub of river cane activity in the country. Inside, Griffith's modest office acts as the trading floor for an informal and evolving exchange of river cane. Beginning in 2004 with David Cozzo, the Revitalization of Traditional Cherokee Artisan Resources (RTCAR) was formed when the Eastern Band decided to support actions to reestablish and grow native crafts. At that time, only three elder women still knew how to make the intricate double weave baskets, known for their complex patterns and their watertight weave. Early on, RTCAR supported an arts teacher in the local high school who knew the skill and introduced it to younger tribal members.

It soon became clear that there was not enough quality plant material to sustain the crafters.

When Griffith was officially hired, he connected with landowners and began his ecological missionary work. When he made a convert, he funded efforts to dig up stands of cane and transplant them to new sites. As word got out about these efforts, landowners who were planning to remove river cane would call up asking if EBCI artisans could use it. If a willing donor site was not too far away, he would find a way to dig up a portion of the cane and transport it.

Once a long-haul trucker from Tallahassee contacted Griffith after a Google search of river cane brought up RTCAR. The trucker said he had some land north of Gatlinburg, Tennessee, that had a lot of cane on it and wondered if Griffith wanted any of it. When Griffith and two basket makers drove to the site, they found the best river cane any of them had seen. The owner thought the stand must be at least forty-five years old. River cane this old can grow to diameters the size of a half-dollar coin with long sections unmarred by the leaf sheaths that clasp younger stems; this is exactly what basket and blow gun makers need. The story seems unlikely; after all, how many truckers are aware of river cane? "I get these sorts of random calls all the time," Griffith says. The interest in this plant seems to transcend cultural and political boundaries. Realizing its utility, people throughout the south are beginning to pay attention to this plant.

Griffith came to the job as well prepared as anyone could for such work. His graduate work focused on understanding the types of soils in which river cane could thrive. He boils down his research this way: river cane isn't choosy, it does very well in sandy soils but also tolerates a mix of construction fill and mulch. It will grow pretty much anywhere except in permanently saturated soils. He has conducted windshield surveys throughout most of the seven western counties of North Carolina to determine where the plant grows. But his current work has taught him that there is still more to know about this plant.

"I've seen transplanted stands sit for five years and do nothing," Griffith explains as we talk in his office, "then it started expanding six feet a year in every direction." There are pictures of river cane on his walls and a box full of harvested cane that has been cleaned and separated into its discrete parts. River cane is one of North America's three native cane species. Functionally a grass, it sprouts from rhizomes that run laterally underground. Similar to quack grass, whose lateral runners ensure that a determined gardener will never really remove all of it, river cane is tough. Yet despite its toughness, the plant occupies only 1 to 2 percent of its historical range in a swath that extends from the Carolina's through Alabama and Tennessee and further west to Oklahoma and Texas. The vast canebrakes that extended for miles along rivers in the south no longer exist. Today evergreen river cane and the deciduous and more diminutive hill cane grow in this area of North Carolina, and the taller "swamp cane" grows further south in Florida.

River cane enthusiasts are concerned about this decline, but the uncertainty surrounding the decline muddies their reaction. It still lurks in ditches, holds on in the strips between rivers and roads, and most commonly hangs precariously along the edges of wet fields where it is regularly pushed back by unsympathetic farmers. The problem is that it's not allowed to thrive. Its ecological benefits of trapping sediment and stabilizing stream banks have likewise vanished. Shepherding existing stands, transplanting patches to other areas, and keeping track of this expanding list of sites is Griffith's job.

Pulling up a PowerPoint developed for one of several talks he is increasingly asked to give, Griffith explains that shortly after European contact river cane was found extensively in the cleared streamside fields cultivated by the Cherokee. Griffith theorizes that the plant rapidly took over these areas when many of the Cherokee were forcibly removed from their cultivated lands. A tragic moment for the Cherokee Nation born from the original sin of our nation, some speculate this might have been the period of the plant's maximum coverage. Written historical accounts don't describe aerial coverage or stem densities. All that is certain is that the plant was used extensively by the Cherokee and many other tribes for a variety of purposes ranging from baskets to mats to blowguns.

Historical coverage is not the only point of uncertainty. The timing of flowering and genetic variability of stands remains a mystery. Most large stands are one plant that has expanded over decades, similar to an aspen grove. A bamboo species in Japan flowered after a seventy-six-year period of vegetative growth, but no one is certain if this example constitutes an outlier or is a baseline. The best guess is that it flowers every thirty to forty years, but no one is keeping tabs on the tens of thousands of small stands hiding out in twenty-three states. Whether this long flowering interval benefits or inhibits genetic health is even less understood.

These unanswered questions aren't stopping Griffith and his expanding network of partners. He instructs grantees to dig up as large a section as logistically possible, taking as much soil and the all-important rhizomes with it. This mass of soil and biomass is placed on a truck and taken to a site where someone wants it.

It is placed on prepared ground, sometimes mulched and occasionally fertilized, and left alone to become established. Griffith has noticed much better survival rates with larger clumps. He also tries to bring cane from multiple stands to new sites. It's a simple process, but it requires a bobcat or a front-end loader.

Finding willing landowners is the long game. Griffith says that land trusts have been a fortuitous partner, but he's also worked with parks departments and private landowners. He fields many calls from landowners who are about to cut down their river cane. This precipitates a search for willing recipient sites. Often the timing doesn't work out. A staff member in charge of stream-restoration projects for one local land trust said there was no river cane available within a reasonable distance when his contractors were ready to plant. The land trust went ahead with a readily available and reliable mix of hardwood trees and vigorous shrubs that would stabilize the site.[3]

Griffith is slowly building a list of successes. "It's more of a marathon that a sprint," he tells me. RTCAR has funded more than fifteen transplanting projects throughout a seven-county area in western North Carolina. His connections are evident when I ask him for background information or contacts. From Oklahoma to South Carolina, Griffith seems to know everyone and has documentation on everything written about river cane in his area. He is a researcher's dream, offering his extensive photo library of hundreds of river cane projects. He demonstrates endless patience when explaining a plant that most people haven't considered. If you want to learn about river cane, talking to Griffith is an excellent place to start.

Despite Griffith's assistance, I still don't understand why this plant has no role in a stream-restoration business that is more active here than anywhere else in the country. North Carolina has a well-developed stream mitigation bank that has facilitated 460 miles of stream restoration or enhancement since 1999.[4] Why is this plant relegated to the edges of our collective human footprint?

STREAM MITIGATION MACHINATIONS

"I pleaded with them to put river cane on the list," Griffith tells me. "I never got a response back." Griffith was spreading the gospel of river cane to the Interagency Review Team (IRT), a group of regulators and reviewers tasked with approving or rejecting the mitigation plans that describe how stream-restoration projects will be implemented in the state. In practice, it is less a species list and more a standard operating procedure that dictates the type of vegetation and habitat that can be planted in stream mitigation projects. The mitigation guidance for stream restoration, developed by the Army Corps of Engineers in concert with the state, includes vegetation "success criteria." In nearly all cases, the accepted success criteria means a forested riparian buffer.

This species composition of "Piedmont/Low-Mountain alluvial forest" is just one aspect of the success criteria. Numbers are critical as well. Specific stem-count targets of surviving plants must be met at year three, five, and seven after restoration. Stem counts must average 210 per acre by year seven and must average ten feet in height, a requirement that pushes for heavy planting of fast-growing trees. Any restoration firm that fails to meet those stem counts will be required to plant additional trees at their cost and do whatever is necessary to ensure their survival. The all-important restoration credits are released by the Army Corps only when the success criteria are met, so there is a huge incentive to plant species that have grown successfully in similar projects. Matt Harrell, a senior project manager for stream-restoration projects in Raleigh, says "regulators want to see a former cow pasture well on its way to a climax hardwood forest. There are no incentives and a few disincentives to doing anything different." Although a restoration firm can technically suggest alternative criteria, a bald cypress slough for example, there is little incentive to do this. As Harrell explains, "Providers typically pick their battles carefully and try to avoid gumming up the works with the regulators, which means that lots of projects come out looking suspiciously similar to the previously approved plans. If it ain't broke. . . ."

For a field that has received criticism over the years due to occasional failures, it makes sense to stick with what you know. The predictable flush of woody species assuages any misgivings about the sacrifice of one stream to development in exchange for the stream system that emerges from the restoration project. The default selection sets the riparian corridor toward a climax state of bottomland hardwood forest—the apex of the potential ecological state—and promises to support all of the biota that could be sustained in this ecological system.

But this mix of hardwood forest community—jokingly referred to as the "mitigation hardwood mix" by many in the business who are keenly attuned to a much wider range of habitat types—doesn't necessarily yield the diversity it promises. Seven years postrestoration, the initial mix of tulip poplar, red maple, sycamore, and live stakes of willows and dogwoods can, in some cases, simplify to forests of one species. "I can take you to fifteen-year-old stream-restoration projects where the vegetation is little more than sycamore trees located eight feet on center," Harrell tells me, annoyed that restorationists can claim a forest when only its most obvious component has been established. The vigorous establishment of one tree species satisfies regulators but ignores the importance of heterogenous habitat on a broader scale. It also precludes certain plants, such as river cane, that don't perform as predictably.

Harrell points to the need for intermediate disturbance as opposed to extreme disturbance that comes with massive forest fires or hurricanes. Intermediate disturbance might involve natural stream migration across a floodplain. It might involve grazing of certain plants by domesticated herbivores. Windstorms that uproot small stands of trees or selected cutting are other examples of irregular but not unusual actions sometimes unintentionally removed when the intent of restoration has been invoked. It can be characterized by a mix of successional states driven by disturbance rather than the stable climax hardwood forest. Although not as predictable, this is the ecological state with the greatest diversity, Harrell explains. Fittingly, there is evidence that river cane is a perfect plant to come in after disturbance. Yet with stream restoration increasingly dominated by the engineering sector, accepting unpredictability is uncomfortable. Unpredictable in flowering and growth rate, river cane doesn't conform to seven-year monitoring windows and stem-count success criteria. It grows when and where it wants. Somehow, a plant that has existed for thousands of years in this area still represents an unnecessary risk.

When I ask what is required to break the logjam, Harrell says that some on the IRT are open to considering the plant but most don't want to rock the boat. A subcommittee has been formed, but no decisions have been made, and key agency reviewers in the Army Corps of Engineers don't want to go off script.

More broadly, Harrell suggests that we need to reevaluate the overly prescriptive language in the permanent conservation easements that dictate management of these riparian areas. "We shouldn't fence ourselves in," he explains. Intermediate disturbance by definition brings with it some uncertainty. It also requires a mix of successional states, some like river cane that might hang on in certain stretches of the river for two decades before transitioning to another habitat type. This condition doesn't necessarily sacrifice performance. Harrell describes one restoration project at nearby Warren Wilson College in which river cane was incorporated into the restoration work. The professors sponsoring the project were happy to use this restoration project as an ecological experiment. "The river bends that were anchored by river cane were the most stable ones in the river," Harrell tells me.

A WILLING AND ABLE SUPPLIER

From Adam Griffith's perspective, there are more immediate challenges. Simply put, there is very little stock to be bought and seeds are scarce. Even if a restoration project wanted to include river cane, he can't point them to a local nursery. A key barrier is that no one knows how to effectively propagate it. One person who met Griffith and heard the call to duty is Thomas Peters. A recent graduate of the University of Georgia landscape architecture program, Peters also pursued a certificate in Native American studies and became obsessed with river cane after meeting Griffith and Cozzo on a trip to Cherokee. Peters sprinkles in references to researchers in the field with disgruntled commentary on certain players. He projects a passionate and unguarded persona, a unique hybrid of a learned academic and a surly member of reality TV shows. His website invites anyone to start up a "caneversation" with him. I am eager to have such a discussion.

Peters has some choice words for the Army Corps of Engineers who recently ponied up $50,000 for a virtual conference that attracted more than two hundred attendees. In decidedly saltier language, he describes the current discussions of the alliance as onanistic and circular: "Acting like they care about indigenous knowledge. And slapping together documents with pictures of Asian bamboo." His comments, although sharp, strike me as less mean spirited and more "discanesolet"; lashing out over a dream deferred.

After seeing Griffith's determined but weighty efforts, Peters became obsessed with collecting river cane seed. The goal was to streamline propagation and provide stock for restoration efforts. This obsession took him to rivers deep in the backwoods of Arkansas where a collegiate friend had discovered a flowering stand. Peters said that they "crossed barbed wire, canoed rivers, and climbed mountains to find the stands of flowering seed." Their efforts paid off, and they gathered "garbage bags of viable seed and brought it back to Athens." Peters started to grow it in easily transportable one-gallon containers with the hope of beginning a nursery operation. "The real surprise" and the thing that set him apart from other growers, as he describes it, was his use of the rhizome material that rapidly grew from the seedlings, as if willed on by the committed cultivator. This "spaghetti noodle" material contained many more nodes per inch. These stringy rhizomes allowed him to start new plants in the high-volume plug trays common to large-scale nursery operations.

Peters started a consulting business on all things river cane and created his website. At the aforementioned virtual conference, he presented on a project of his at the Cowpen's National Historic Battlefield where he helped the National Park Service re-create a historically accurate landscape by seeding an area next to the site where British soldiers were served their comeuppance. Peters claims he spent ten years figuring out how to grow this plant and, as if holding onto some stock options of uncertain value, is admittedly guarded about describing the process. Peters grew plants with an eye toward expansion, but a typical order from a nursery might amount to only four or five plants. He tried to match the prices of other stock but says, "given the time required, this plant has to cost more than a pine seedling. Maybe that's where I went wrong."

Today Peters has taken a job managing a "greenwashed private community." At the end of the day, cane doesn't pay the bills. He hasn't yet been able to introduce river cane onto that suburban landscape. When I ask him if river cane is still in his future, he says, "that's the dream!" "There's a lot of lip service toward river cane," Peters says, apparently done with talking and ready to jump into the trenches. Wistfully he suggests that if even half of those $50,000 of federal largess were applied to a contract to produce river cane, his dream of providing the plants for restoration efforts in the southeast could become a reality.

OLD MOON, NIGHTHAWK CHEROKEE CANE

In my investigations of river cane, I came across a lot of learned enthusiasts, however, for what it's worth, mostly of the white male variety. My head was swimming in the minutia of plant lists, soil types, and heavy machinery. I was seeking a bigger picture of what this plant has meant for rivers and what it could mean. I found what I was looking for in Roger Cain, whose relationship was both personal and homophonic. His take on the plant spanned centuries and cultures.

"You can't write a book about rivers without writing about cane," Cain tells me soon after we meet. He speaks with a soft western drawl of the plains and a straightforward manner that is disarming. Cain is part of the United Keetoowah Band of Cherokee (UKB), a smaller band that moved west, initially to Arkansas and ultimately to Oklahoma, before the forced removal and the infamous Trail of Tears. Half Cherokee and half white, Cain speaks of his Cherokee side as the "old settler and Nighthawk Cherokee." He relates a complicated history that involves bad blood between the Keetoowahs and the main western Cherokee nation. The UKB's modest land holdings are located within the much larger western Cherokee lands. The band is currently unable to build its own casino, cutting off a stable source of revenue. After applying for recognition in the 1950s, they were eventually granted recognition by Nixon ("only good thing he did"). Speaking to Cain, I am reminded of the multifaceted conflict that oozes out from the original sins of colonization and forced removal of a people.

Cain had just returned from a trip to a six-year-old project where Peters's cultivated river cane had been planted. Cain was attempting to burn the stand to stimulate future growth, and I ask him how it went. "Didn't burn," he says, it had been a wet spring in eastern Oklahoma. I ask him about a suggestion he made at a recent conference to burn with the "old moon." Cain explained how cane can act like trees where the water is held up in the culms during full moons. "You can hear it sloshing up there," he says, offhandedly providing an image that stops me in my tracks. Toward the end of a waning moon, Cain says, the culms are drier and should burn better. I consider the science but also hold on to this luminous image of water-laden river cane swaying in the wind under the power of a senescent moon. I envision the moon holding water, its gravitational fingers reaching down, imperceptibly, to the rhizome network of this native grass.

Cain grew up playing in canebrakes on Cane Creek. I wonder if the recurring iterations of the word "cane" are coincidence or some sort of verbal catnip used to entice a writer like one might excite a bird dog with pheasant feathers. Regardless, I'm as focused and expectant as a spaniel awaiting the hunt. Cain explains that his father was a "medicine-man" who shared his knowledge about plants and the Cherokee relationship to them. "River cane is part of our cosmology—we use it in the holy water ceremony." He trails off, realizing that this lesson might be too advanced for his audience.

Cain says that his interest in river cane was renewed about fifteen years ago when, then a high school math teacher, he began teaching some students about stickball. One thing led to another, and eventually he attended graduate school to study ethnobotany. He is currently employed by his tribe as an ethnobotanist.

As Cain speaks in free-form arcs about river cane, I'm struck that his reference points jump back and forth from the white perspective to the Cherokee one.

The cattle industry has river cane to thank for its existence, Cain tells me. It was the original source of food for the cattle the early settlers brought along. "Cane provided shelter and food—and cattle came out fatter. That's what helped tame the wild west," he explains. In yet another example of betrayal, he says, when land became scarcer and people started to cultivate the land, they "turned on the plant." Farmers began to mow it down and tear it out. People say it's invasive. "That's bull."

I ask what role river cane plays for the United Keetoowah Band today. There are a few weavers, his wife being one, but he describes that obtaining the plant is fraught with issues. Long-term relationships with landowners don't usually continue with the younger generations or new owners. There are efforts at reciprocity. Every couple of years a basket maker might give the landowner a basket in appreciation. Sometimes they might pay by the culm. Cain seems reluctant to discuss these oftentimes thorny transactions. "Sometimes river cane is treated just as a commodity. I like to think of it more holistically." Between the wildlife it supports, the carbon sink it offers, and the improvement to water quality it provides, the plant is more than its culm. Even our understanding of its range may be limited. Cain explains how he recently sent a culm to a person from the Navajo Tribe who wanted to make a flute, similar to the one seen in the ubiquitous images of the fertility deity Kokopelli. "That means river cane was likely as far west as Arizona," he explains, expanding the known range of the plant a thousand miles westward, its water-filled leaves swaying in the desert wind.

ASK THE PROFESSOR . . .

An air of mystery seems to permeate all aspects of this plant. A slightly different interpretation emerges from every person with whom I speak. The confluence of a beloved and imperiled craft with an overlooked and besieged native grass pushes people to make statements that span centuries and divergent cultural paradigms. As the question marks compound, I wonder what is really known about this plant. No one seems to know how frequently this plant flowers, how much currently exists, or the area this plant covered at any particular time period. People have differing opinions on even basic questions such as how much sunlight it prefers.

I reach out to an academic who has spent more time trying to figure out where this plant existed than anyone else. Bruce Hoagland is a professor at the University of Oklahoma who spends half of his time coordinating the Oklahoma Natural Heritage Inventory, the definitive compilation of plants in the state. He lists as one of his research interests the reconstruction of historic vegetation. With his paisley shirt, longish hair, and handlebar mustache, Hoagland appears to be an isolated member of his population who has taken root in a pocket of acceptable habitat in central Oklahoma. Clearly a plant geek, a subculture that can come across as insular to outsiders, Hoagland is nevertheless happy to share whatever he knows.

In his mellow Oklahoma twang, Hoagland tells me that he and a long list of students have tried to figure out where river cane existed and how much was present. The primary source of information is General Land Office surveys conducted in Oklahoma in the 1870s in the western half of the state and in the 1890s in the eastern half. Although the surveyors would note canebrakes when their survey transects ran through or near them, delineating vegetation types was not the focus of their survey. Glimpses into the historic vegetation of the area can be gleaned from summary notes they compiled. These offhand observations, written in barely legible cursive, provide a qualitative picture of the vegetation types present at that time. Hoagland says he would love to have a student conduct a comprehensive review of these survey notes and develop a more quantitative assessment.

Hoagland is pleasantly surprised by the fervent interest in river cane these days. It's clear that he's not used to people being so interested in plants. During his plant surveys for the Natural Heritage Inventory, the most common response he gets from regulation-leery landowners is, "I hope you find something interesting, and I hope I don't find out about it!" His connection with river cane extends back to his childhood in Kentucky where he recalls fishing trips to his grandfather's rented shack on the Salt River. "We used to actually fish with cane poles," he says, not unlike a hipster who is surprised that everyone else finally discovered his favorite band.

When I try to confirm Cain's theory of river cane extending to Arizona, Hoagland tactfully says there has been "a lot of brainstorming" related to the plant. He says the current patchy distribution is definitely the result of postcolonial settling and development. But how far the canebrakes expanded is harder to tell. He points to Thomas Nuttall's 1819 account of a brake extending at least a square mile into the current town of Three Rivers, Oklahoma. River cane likely made up a significant component of the larger bottomland hardwood forest ecosystem. Confirming Griffith's explanation, Hoagland points out that river cane won't grow in the most saturated soils. With riverine bottomlands possessing a great topographic diversity, the coverage of river cane was probably limited to those areas where it could thrive. I ask him if this means that riparian areas used to be more diverse than they are now. "We have a tendency to say diversity is the goal to achieve. But in some situations there wasn't diversity present."

When I ask Hoagland about the claim that river cane occupies only 1 to 2 percent of its historic coverage, he doesn't disagree. "It's no doubt that the current extent is in some one-digit percentage of what was here prior to European settlement." The biggest upside of this renewed interest, he says, is that it might start bringing people together to talk. The tribes are often suspicious of each other as well as anyone with a government association, so no one really knows what anyone else is doing. Hoagland seems happy to help should the scattered discussions develop into plans that might require his expertise.

DYLAN MORGAN: BASKET MAKER AND CANE WEAVER

I was relieved that one basket maker was willing to talk with me. My calls to a few of the elder women were not answered. Perhaps they were a bit tired of all the press surrounding basket making. Not long ago, these were the last three basket makers of the Eastern Band of Cherokee Indians (ECBI). It's a story that hits an evocative minor chord with my culture—the last remaining basket makers of the (ECBI) quietly doing their craft. It's novel, nonthreatening, and stimulates our consumerist desires. "Maybe I should pick up one of those," we think as we peck in a Google search for the going rates.

I wanted to understand the craft that this plant sustained. Although many other tribes used river cane (as well as oak and honeysuckle), the (ECBI) has been able to sustain the craft thanks to a few elder artists and a thriving tourist industry that draws fourteen million visitors annually to the Smoky Mountains. A prized, double weave basket made by an established artisan can easily sell for more than a thousand dollars. A basket made by a well-known elder recently fetched a tidy $75,000.

Dylan Morgan was happy to talk. Fresh off an opening event at the Center for Craft in Asheville, where his work was showcased alongside that of eight seasoned Cherokee artisans, he wasn't surprised by my interest. We headed to a Cherokee-owned coffee shop that Morgan recommended. Our drive took us past a forty-foot-high nonnative bamboo stand that stood incongruously on an island in the Oconaluftee River that flows through downtown Cherokee. The bamboo had been planted to stabilize the stream banks, and it now formed a forest that could sustain a family of pandas. I mentioned the irony of this exotic interloper flourishing where river cane might grow, but it didn't register to Morgan as odd because it had been there for his entire life.

The town of Cherokee is a study in contrasts. An attractive and informative tribal welcome center sits down the road from the Gold and Ruby Mine tourist attraction. On its sequined sign, a white-bearded prospector armed with a pick and a side-kick burro implores passersby to "pan fer gold!" Not far from Qualla Arts and Crafts, the nation's oldest Native American fine arts cooperative, is the Totem Pole Craft Shop, which supplies knives, leather goods, and dream catchers to keep the kids occupied on the eleven-hour drive back to Wisconsin. Native Cloud CBD, a vaporizer store, promises relief for the parents on that same journey.

We get some coffee and sit outside on a deck overlooking the river. Morgan left a small, half-finished basket in the car. With his long black hair pulled back, steel tunnels in his earlobes, and tattoos adorning his arms, in another context I might have guessed that he just came from a punk show rather than from the Museum of the Cherokee Indian where he works. As he regularly employed his vaping stick, I wasn't sure if we'd have anything in common.

I began by asking about a trip he took with our now mutual acquaintance, Adam Griffith. I asked about the stand of legacy cane from the long-haul trucker's land in Tennessee. Morgan confirms the score. "It was some of the best I've worked with. Usually, I might need twenty canes to produce a medium-sized basket. With this stuff, I needed only four or five canes." This wasn't his first long-haul trip to find good river cane. Morgan took a trip to Kentucky with some of the elders and woke up in the parking lot of a Radio Shack. The store manager, who was also some kind of town official, produced an original signed document stating that the Cherokee could harvest river cane in their town "as long as relations remained friendly." It wasn't clear if this was authentic or even necessary, but they were happy to harvest some cane. Morgan laughs at the cultural dissonance.

Some harvesting trips closer to home weren't so agreeable, and Morgan describes how one old-timer notified his group that they "had three days to get over here and take it." On other gathering trips where they had secured permission from the landowner, he felt the suspicious eyes of neighbors boring into them. He shakes his head over this generosity tinged with hostility.

Unfortunately, there aren't many locations within the Qualla Boundary—the local term for the lands held by the (ECBI)—where large stands of river cane can be grown. Some parcels further south owned by tribal members have some decent river cane, but this is also a drive. Two years ago Morgan heard from some National Park Service staff that they might allow his small group of weavers to plant and harvest patches in the nearby Smoky National Park. "That would be fantastic—being able to get into the park for free and having a large stand." Morgan hasn't heard anything else about that idea and is unsure whether the pandemic or bureaucracy stopped the idea.

Knowing that until recently only three elder women carried on the craft, I ask Morgan about the gender role that seems to come with basket making. "Yea, some people give me crap about that." But then he runs through the multiple steps: searching for cane; gathering, stripping, and preparing the strips; gathering dyeing materials and the dyeing itself; and finally, the weaving. "If that's women's work, it's more work than whatever they're doing." Morgan waves off this irritation as well as he begins to speak of his teachers. He first started as a wood-carver. "I quickly realized that my brother had the real talent for that," he says, smiling. He started to hang around Lucile Lossiah and Ramona Lossie, elders who engaged him with small tasks. "They would ask me to prepare some strips or something like that. And we'd get to talking."

As Morgan tells it, his training in basket making began slowly and evolved as he started to hang out with the weavers. He wasn't the only young person. He mentions a younger tribal member, Gabe Crow, who he describes as being both helpful and further along with his craft. He jokes about how they call Lucile the GOAT of basket making. Beyond the humor, the respect is clear. He feels fortunate to learn from such accomplished women, while also having the comradery of a small group of three or four younger weavers.

Currently he's receiving attention for his craft. His baskets and mats are being displayed at a show in Asheville, and occasional traveling exhibits are organized by the local museum (figure 7.2). He has made some custom orders, but Morgan doesn't want to be "filling orders." After demonstrating the craft at the museum all day, he wants to weave the projects he wants to weave. Producing baskets for steady contract orders would rob the work of a certain joy he draws from it. The

7.2 Dylan Morgan's river cane snake mat, made using bloodroot and walnut dyes, on display at the Center for Craft in Ashville, North Carolina.

basket work and the connection it provides run deeper than being a source of extra spending money. He is determined not to lose that. I sense a restless energy but also a connection that grounds Morgan. He tells me at least ten times about his respect for the elders. When talking about the museum where he works as an interpreter, he describes it as great—but it could be "so much better." Now a full-time employee, Morgan seems to be angling to shape some modernizations and to have an opportunity to tell a fuller story of his people.

Morgan also chafes against some of the expectations of his role as a tribal member. He relates how some people give him sideways looks for seeking information from Adam Griffith, a white guy with a PhD. "I tell them he knows a lot about river cane!" As the recorder and scheduler of harvesting activity, Griffith plays a central but potentially awkward role in maintaining the sustainability of the harvest. Harvesting sustainably is important, Morgan says, but it can be a challenge because of the randomly located stands used by a growing number of artisans. And river cane is just one component; a number of local plants are also needed for dyeing. Although he's heard of some people stripping walnut bark from living trees to achieve a dark brown, he uses the fallen hulls. "They are surprised I get a deep brown, but I tell them I boil it for two days."

I ask how the pandemic has affected his band, and he explains that the tribe took strong precautions to protect the elders. They even shut down all access into and through the Qualla Boundary for more than two months. "It was all about protecting the elders, protecting the native speakers," he says. "I wasn't aware we could even do that," he continues, pointing to a collective action that seemed to evade the rest of the country. As we speak, Morgan weaves strands of loss into our discussion. Despite his tribe's relative success against the virus, his immediate family suffered other ills. His mother came down with cancer a year ago. He returned home from Albuquerque where he was studying engineering. College wasn't a good fit and it was important to be back with his family. Shortly after his mother passed, his father and grandmother passed as well. In just a year, he lost three of his immediate family members.

I recognize the pattern, made fuller by the dark walnut stain that contrasts with the sunnier yellow. Death and acceptance, reflection and newfound purpose, bound tightly into a form that provides both utility and beauty. Although he doesn't need my help, I'm pulling for him to succeed. With elders providing guidance and his peers providing inspiration, the only thing that is missing are adequate quantities of river cane. The power of this connection, even to someone who lies outside of its cultural context, is palpable. It lies just under the surface, waiting to run.

HOW IMPORTANT IS OUR RELATIONSHIP TO A PLANT?

One of the first questions people bring up when discussing restoration is, "What are we restoring to?" It's an important question that nearly always brings other questions with it. A provocative conversation starter, but it rarely results in a settled conclusion.

I wonder if the question isn't better phrased as "for whom are we restoring?" In Robin Wall Kimmerer's book *Braiding Sweetgrass*, she describes a research project undertaken by her student, Laurie, on the effects of harvesting on sweetgrass growth and vigor.[5] The native practice of taking no more than half of any stand is part of the "teachings of grass, *Mishkos Kenomagwen*," that are followed by the Anishinaabe. Kimmerer's student sets out to test this native wisdom against the rigors of the scientific method and constructs an experiment to determine what impact harvesting has on the plant. The proposal is met with dismissal from the academic review board because "anyone knows that harvesting a plant will damage the population. You're wasting your time." After a rigorous experiment that mimics traditional harvesting practices, the student finds that the practice does result in increased growth and vigor in the plant, whereas the unharvested control plots were choked with dead stems. The student points to several analogues that inform what she has seen: the beneficial effects of fire, and the stimulative effects of buffalo grazing on grass. She points to the physiological characteristics of grass stems, whereby the severing of a blade at the base can stimulate numerous nodes that will sprout new stems that can access the newly available resource of light.

Kimmerer explains how her student's research wins over the previously skeptical review board and validates traditional knowledge through the lens of the scientific method. But this epiphany wasn't needed by the elder whose harvesting practices framed the study. The elder, Lena, describes the traditional knowledge succinctly: "If we use a plant respectfully, it will flourish. If we ignore it, it will go away." This role of an engaged and conscientious use, distinct from passive appreciation or recreational use, is a missing element in most river restoration efforts.

With this giant southern grass, river cane, we have a small but unique opportunity to reestablish a more regular relationship. For many southeastern tribes, this relationship is both dateless in its roots and timely in the current moment of reconnection. For the basket makers, the immediate connection includes the more tangible decisions of who can harvest and what specific craft will emerge from it. For those not involved directly in the craft, there are secondary relationships that might flow from being connected to this craft.

Landowners checking on a cane stand for a favored weaver might observe a riverbank that is newly exposed and unstable. They might observe that algae now covers the stream in June rather than later in the summer. They might notice that a particular species of fish is no longer present. Whatever they observe, it would be grounded in a direct connection, imbued with personal relationships and tangible products. It would move beyond the realm of regulatory credits and self-constructed ideas of purity. The state of our rivers might become real—the equivalent of the kitchen table issues that so frequently determine our elections.

What would be the point of these granular observations and heightened awareness? I believe we only truly engage when we can see a constructive role we can personally undertake. A growing number of stream technicians can design a restoration project that will accommodate the physical forces of the stream involved. But only a few restoration projects engage people in a recurring relationship. How might our collective resolve change if we had this relationship? Would we be more likely to support policies that restrict development to limited areas? Would we accept restrictions on routine actions such as salt application if we physically measured the acute toxicity overapplication creates in our rivers?

For most of us, it is far too easy to ignore our rivers. We are not dependent on them in immediate and tangible ways. Collectively, most of us have no role in our stream-restoration projects. They don't demand our regular attention, and our attention predictably drifts elsewhere. Perhaps the only way to cut through these distractions is by creating connections that bind us with the resource. Like the long-haul trucker who knew he had something unique on his Tennessee property, we can seek out those who use this resource and make it available. Whether given freely or traded, a connection is made. Our understanding of the river is rooted. An understanding based on use need not be exploitative. With respectful use and the attention that comes with it, we have a stake in the survival of our rivers. Through this respectful use, we might ensure that they flourish on our watch.

CHAPTER 8

NATURALIZED CHANNELS IN MILWAUKEE

Removing Concrete and Lowering Floodwaters

UNDERWOOD CREEK: ANSWERING THE CHALLENGE

I meet Tom Chapman at Underwood Creek, a tributary that drains into the Menomonee River that in turn flows into Lake Michigan at the Milwaukee Harbor. He's wearing a windbreaker and the standard office casual get up. Chapman bears a passing resemblance to Steve McQueen that is enhanced by his cool demeanor. Wauwatosa is a modest but desirable neighborhood of bungalows with tidy lawns and the occasional Brewers flag. The creek runs along a CSX rail line beyond which lies a large section of public land with a mix of fields and forests known locally as "County Grounds." Upstream, a north-south arterial highway rises above the floodplain to the height of the prominent Wisconsin Steel and Tube sign, and the low drone of rush hour traffic fades into the background. Most people would find the neighborhood and its mix of residential and light commercial uses pleasant but not particularly noteworthy. At first glance, if anything big happened in Wauwatosa, it must have happened some time ago.

Chapman and I step away from our cars and walk down the gentle slope toward the stream. Ten-foot-tall maples and river birches are just leafing out, and their protective brown paper tree wrap remains around their trunks. Golden Alexander plants add a fuzzy yellow haze to large swaths of the floodplain. The stream, about six inches deep today, winds through an assortment of large boulders in the channel. Two years after the completion date, it's becoming difficult to tell that this was the site of a major restoration project. Large willow trees close to the channel remain, spared during the extensive earthmoving. Unlike the early successional uniformity of the vegetation surrounding most stream-restoration projects, this landscape is more difficult to date.

Chapman and I have known each other for a couple of years. He helped me out in my work with the Milwaukee Metropolitan Sewerage District (MMSD). When I was running a class for MMSD, he delivered an in-depth presentation on the sewer and stream history of Milwaukee. For some planning work I've done, Chapman has provided useful summaries and maps and clarified certain issues for me. He's not exactly a client nor exactly a colleague. I called him here because we're on friendly terms and I know he has worked on these types of stream projects for more than twenty-five years. His greeting this afternoon, neither unfriendly nor encouraging, is a straightforward "So what are we doing here today, Pete?"

The project in front of us is one of MMSD's most impressive stream projects to date, and it's difficult for Chapman to not talk about it. Three years ago, this creek ran through a trapezoidal concrete channel devoid of all life. It was a barrier to any fish or aquatic life because it normally ran only an inch deep. With any significant rain, the shallow trickle rapidly swelled to a violent torrent. Today the water runs deep enough for a Coho to swim up, and the water is clear enough to see the cobble that now lines the bed. A mother duck herds a surprisingly large flock of a dozen or so ducklings downstream. Red-wing blackbirds sound their familiar shriek, and four wild turkeys are gathered along the bank downstream with uncertain purpose. I did not see the channel before its transformation, so the scene before me seems normal and right. With the bounty of wildlife and warmer spring weather, the assembled verdancy is a gift from the gods after the punishment of a Wisconsin winter. Dispelling me of the notion of any divine intervention, Chapman reminds me that the aquatic rebirth in front of us resulted from the hard work of many over five years and an investment of $11.6 million from MMSD and their partners.

Chapman moved from his position as the manager of the section that oversees all stream projects to become a special project manager for the director about three years ago. MMSD created a new position for the experienced staff who had been in one position for some time and were looking for new challenges. As a "senior fellow," Chapman handles a range of tasks, from providing exhaustively researched PowerPoint presentations on the history of water and sewer conveyance in Milwaukee to attending public meetings on MMSD's watercourse investments. As a sort of agency consigliere, he frequently vets potential projects and approaches before they reach the leadership. This sort of vetting is an initial step for projects that require years of planning and partnership building. Much can go awry, and part of Chapman's job is to keep tabs on this.

This role keeps him busy as MMSD is currently pursuing a stream-restoration goal that is startling in its ambition. Seventeen miles of concrete channel are located in Milwaukee County, and MMSD intends to remove fifteen miles of it. As of 2018, they had spent $432 million removing concrete-lined channels, and they intend to spend another $533 million going forward. These numbers include a range of related actions, such as property acquisition, creation of flood detention basins, and stormwater management, which are integral to the more visible work in the channel. These numbers would seem large in major cities like New York and Los Angeles, and in the Midwest, they are eye-popping.

I ask him what it's like to work for a boss who continuously throws ambitious new goals to his staff. He sizes me up, wary of my angle. "Kevin always has something new," he states in a neutral tone. In his previous position, he always knew what would be on his plate, but that's not the case anymore. He was recently sent on a mission to investigate whether the introduction of more beavers in the upper Milwaukee watershed might help reduce flooding, limit CSOs, and help address other water quality problems downstream. Reluctant to reveal that he might be having fun, he hedges a bit about the potential for a "beaver solution." "I'm not convinced that its worth flooding farm fields and pissing off a bunch of farmers," Chapman says. But the enthusiasm with which he strays off topic reveals that he enjoyed the thought experiment. The role of beavers in flood management was not part of his civil engineering coursework. Pushing my luck, I ask if it's difficult to balance the constraints of these projects with the big picture goals. "I don't mind the big goals," he says. "I understand what he's doing, always pushing us to do better." He ruminates on the question a bit, and I see his neutral expression soften imperceptibly to the suggestion of a smile. Finally, he nods and simply adds, "I think it's good."

———◆———

As we wrap up the walk, I ask him how involved he was in this project. He wasn't the project manager for MMSD, and the Army Corps of Engineers handled the design, permitting, and contracting. Although I'm not an engineer, I suggest that the engineers came up with an enlightened design on this project. He starts on a meandering riff about how he doesn't like it when people say "they're not engineers, but . . ." "It devalues their role before any ideas come out. The fish biologists are just as important in a project like this." Knowing

Chapman and his dramatic range, this is uncharacteristic and combative by Midwest standards. "It puts engineers on a pedestal—one I don't think they should be on."

I ruminate on Chapman's broadside and realize that my off-hand comment about not being an engineer was a regular saying of my former boss that I've somehow internalized. His contrarian take on the primacy of engineers is energizing and could be considered blasphemous in certain halls of the MMSD's central office building. I didn't take Chapman for the rebellious type. We look upstream toward the highway overpasses that safely carry thousands of people across the newly created floodplain every day. No one is around to argue for the old guard of engineers. I mull over the similar tensions that play out in different streams managed by different people. Highway noise hums in the background, and swallows swoop across the channel. The stream gurgles around the randomly located boulders that grace the channel, and the turkeys continue their uncertain flocking.

I think back to Chapman's initial question: "What *are* we doing here today?" Neither of us are managers of this project, but both of us are keenly interested in the stream in front of us. We are witnessing the evolution of a stream from something straight, hard, and ignored to something sinuous, alive, and valued. It's not complete, but it's rapidly developing as an ecological beachhead in this otherwise tamed suburb. If I had to sum it up, I might double down on Chapman's simple but true assessment and say that what's happening here is more than good—it is transformational.

KEVIN SHAFER AND MMSD: BATTLING FLOODS AND THE REBIRTH OF A SEWER CULTURE

Sewer and water departments aren't supposed to be cool. Like trash pickup and disposal services, we want someone else to deal with the service. With sewer systems, the process conveniently takes place underground. It is subterranean in mind and body, forgotten and disregarded as soon as our unwanted waste leaves our sinks or toilets. It only rises in our attention when there is a problem. This is frequently accompanied by rage toward something we took for granted. When that happens, we would rather not deal with the problem. For nearly all of us, the vast network of pipes that bring us potable water and handles our deluge of waste falls under the category "not my job."

It might make sense to keep their work under the radar and underground, so it is notable that MMSD has made concerted efforts to bring their work into the light of day. Over the past decade, MMSD has become the regional leader in any effort related to water quality, toxics remediation, river restoration, energy sustainability, green infrastructure, and land conservation. In the process, general knowledge about the mechanics of water and sewer movement has been raised in unique ways. A former pump house of the nineteenth-century version of MMSD has been converted into a busy coffeehouse in the city. The now cleaned up pumps testify to a former era when water was pumped from the lake to flush the Milwaukee River of its filth. It would not be surprising to hear normal people reference Milwaukee's proud past as the city of sewer socialists, harkening to a period from the 1880s to the 1910s when Milwaukee had an unlikely run of socialist mayors. In this process, this city's sewer and water department has become surprisingly cool.

When you enter MMSD offices, the walls of their modest lobby are covered with a wall to ceiling panorama of Lake Michigan in the summer. It makes a positive first impression. This is just one part of a savvy public relations effort that includes an attractive website with maps showing how the neighborhoods and streams have changed over time. Away from the offices and out in the neighborhoods, evangelistic staff attend public meetings, and consultants continue the messaging. They tout the numerous MMSD programs that benefit the local streams that feed into Lake Michigan. This change has steadily developed in the past decade. For a city known for its advanced sewer system in the 1800s, it's a return to a familiar story. For a municipal governmental agency, it's a return to relevance.

The individual in the office who is churning out these ambitious goals initially resisted the city he now calls home, and the city now embraces him at every turn. Kevin Shafer, executive director for the past twenty-three years, grew up in Rantoul, Illinois, a small village in east-central Illinois whose only real claim to fame was an Air Force base that closed in 1993. His mother and father worked in school administration and supplied the local Air Force base, respectively. Immediately outside this town of 12,000, corn and soybeans covered the landscape in 160-acre sections as far as the eye could see, creating a world of right angles that might seem fitting for a future engineer. For college, Shafer made the familiar fifteen-mile journey to Champagne-Urbana—home of the University of Illinois. Four years later he left with a wife and a degree in civil engineering.

If this beginning seems particularly wholesome in a midwestern way, perhaps that is why heading north to Milwaukee never appealed to him. At the time, Milwaukee primarily excelled in cold winters and heavy German food. It was a town past its prime, saddled with a legacy of racial segregation and defined by an alcohol-dimmed culture lionized by Mr. Baseball, Bob Uecker. Shafer headed in the opposite direction, south to Texas, where his brother provided a bed to crash on while Shafer started a job with the U.S. Army Corps of Engineers. This steadiest of steady gigs offered a tuition reimbursement program that he took advantage of to get his master's degree in engineering. At the Corps, he learned the standard conveyance designs the Corps has inflicted on the rivers and streams across the country. He enjoyed the stress free, nonbillable environment of the Corps. Lodged in his comfortable abode of engineers quietly calculating peak flows for trapezoidal channels, Shafer had no idea that he would be leading the charge to remove these structures twenty years later.

When Shafer moved back to Chicago, he worked at a major consulting firm for several years. This job brought him up to Milwaukee for certain projects, and this work introduced him to some of the leaders of MMSD at that time. These connections developed, and Shafer was recruited to be the director of technical services for MMSD in 1998. It was a pivotal time. In the previous two years, the Milwaukee region had experienced three storm events in which more than eight inches of rain had fallen in less than twenty-four hours. In addition to widespread property damage, twelve children had died, having been caught up in the floodwaters. Shafer says that this unfortunate incident put him in the business of dealing with stormwater.

Agency morale had been beaten down as well. "The culture did not revolve around pride and certainly not around ambitious environmental goals," Shafer said. At social outings, many MMSD staff would say they worked at the local McDonald's rather than the local sewer and water authority. The ripe odor from the Jones Island sewer plant sometimes wafts over the city and seemed to follow workers into their homes and private lives.

In his first year, Shafer commuted back and forth from Chicago. Given the hostile reception, one could imagine that he fondly remembered the tidy calculations of channel flow back in sunny Texas. But this itinerant arrangement could not continue, and his boss told him that he was needed in Milwaukee. The combination of biblical floods and a hostile public could not be tended to from afar. The city proper was only one part of his restless audience. Milwaukee County, the approximate boundary of the service area for MMSD, represents the full spectrum of political leanings found in the country. There are militantly liberal villages pressing for ecological sustainability and aggressively conservative towns still tarnished by the white flight of the 1950s. Despite these divergent political views, everyone could agree that flooding was bad and that MMSD wasn't doing enough about it.

THE EVOLUTION OF MMSD's STREAM-RESTORATION PROJECTS

Through Wisconsin state statutes, MMSD had been given the authority to manage the streams and rivers in its service area. The initial intent was to prevent overbank flooding that could adversely affect wastewater infrastructure. These impacts could include the destruction of conveyance pipes through erosion or costly and unwanted infiltration of sewer pipes from floodwaters. To the extent that homes flooded and residential water and sewer infrastructure became damaged, that also became MMSD's problem. From the 1940s to the 1970s, MMSD implemented many stream bank stabilization and conveyance projects. These projects were common at that time and converted most streams into open concrete trapezoidal channels, termed *watercourses* by MMSD staff. These drainage channels served their purpose for roughly half a century, but unregulated development in the suburbs laid down thousands of acres of additional impervious surface and exacerbated already flashy flows. By the 1990s, many sections of the channels were filled with sand, stone, and willows that sprouted through cracks in the concrete. Although the solution was not clear, everyone knew something needed to be done. No one, especially the engineers who created them, liked seeing these channels choked with debris and overtaken by opportunistic vegetation.

It all started at Lincoln Creek, a completely urbanized tributary to the Milwaukee River. The stream ran through northwest Milwaukee, a sprawling section of the city historically home to many Black neighborhoods. More than 1,600 homes sat in the hundred-year floodplain in this watershed. Most were single family bungalows and duplexes that housed a population who had worked in numerous factories prior to their closing in the 1980s. Multiple planning efforts were undertaken that identified different approaches. In many similar situations, the most cost-effective solution was to purchase and raze properties at risk.

Shafer describes this as a challenging moment. There were only two staff members tasked with this type of work at the time, and "they were getting the snot kicked out of them." City leaders also weighed in with their constraints. Dave Fowler, a longtime MMSD employee and one of these two staff members, revealed that the local alderperson came to all of the meetings stating that no homes could be razed. The reduction of tax rolls and the tactical retreat of a neighborhood would not be tolerated by local politicians. MMSD would need to find another way.

MMSD did find a way, but it was not cheap. Focusing on the reduction of flood risk allowed MMSD to tap sizable utility revenue funds from the municipalities for stream work. The first portion of the Lincoln Creek project involved digging an eleven-acre wetland in a lightly used state forest that abutted the creek. This wetland, located within the city, retained ninety-acre feet of water (equivalent of a foot of water covering ninety acres) upstream of the flooded area. Further down, the channel was widened and lowered, culverts were enlarged, and cobble replaced concrete in the stream. Vegetation was planted and allowed to grow on the banks, and the stream once again resembled a stream. Three properties were eventually purchased, at above market rates, Shafer pointed out. Five years later, after a $117 million price tag, the project was complete. By altering the hydrology and lowering the stream, the flood maps were effectively changed, and 2,025 properties were now protected from the hundred-year flood.

Despite meeting this impressive goal, this first project was a trial run for MMSD's new approach. All MMSD staff I have spoken to quickly state that "this isn't how they'd build it now." During a visit to the stream, I noticed that the stream remains firmly under human control and influence. Like a dog with an antichew collar, it is a glum and constrained vision of stream restoration. Choked with litter and placed at the bottom of a deep valley of mowed turf, the stream runs efficiently through a narrow band of scrappy trees and shrubs. Although it is accessible, no one would walk down the hill for a streamside stroll. There are places where an in-stream structure still maintains a pool, but the impact of stormwater on this stream is evident from the wrappers and leaves stuck in trees six feet above the water line. Although an improvement over a concrete culvert, it is difficult to envision this stream sustaining anything more than an anemic ecological community.

Considering the impacts over the past decade, the removal of homes from flood risk had been a major success. Although this approach seems to make common sense, it isn't standard practice in this country. When floods and hurricanes damage homes, the most common practice is for people to build back. Originally designed to discourage building in floodplains, the National Flood Insurance program subsidized by the federal government has in many ways enabled communities to continue to build there. Equipped with insurance, people have become brazen in their acceptance of risk. But this risk is not shared equally by all U.S. taxpayers: 94 percent of the public are paying for the 6 percent who live in the floodplain.

Occasionally homes are elevated, as was the case in Mandeville, Louisiana, where more than 80 percent of the homes were raised at taxpayer expense after Hurricane Katrina flooded most of the town. Accounting for this repair work gets fuzzy. Communities are discouraged from allowing development in these areas because of the prescribed cost-share formulas that hold local communities responsible for part of the repair costs. In practice, these cost-share requirements are frequently waived in the wake of major disasters by politicians looking to save the day. The ongoing danger and costs associated with this don't seem to penetrate our collective consciousness. Despite these regular disasters, one website focused on top relocation locations ranked Mandeville in the top 100 places to live in the country. Seen in this light, MMSD's focus on removing the risk to homes and reducing the potential for costly repairs is enlightened.

MMSD's approach evolved with Hart Park, a $48 million project completed in 2007 that addressed a decaying section of concrete channel that ran through downtown Wauwatosa. Just a couple of miles downstream from Underwood Creek, this area is central to a community that has centered its commerce around the Menomonee River. In the trendy, walkable "village" portion of Wauwatosa, new bars and restaurants straddle the river. For Hart Park, MMSD purchased approximately twenty properties in the hundred-year floodplain and expanded the footprint of the park by thirty acres. In addition to removing concrete channels and naturalizing it with boulders, MMSD created new park amenities where many of the homes had existed. MMSD reduced the extent of the floodplain by lowering the land where the park would be located by three feet. The playgrounds, skate parks, and playing fields included in the project would be able to withstand the eventuality of future floods. Fowler, now a floodplain manager for FEMA, says this demonstrated how a natural flooding solution could bring significant additional value to a community. Although he alludes to some complaints that this project provided benefits not seen in the Lincoln Creek project, he maintains that this has been a natural evolution for MMSD.

CULTIVATING COMMUNITY ON THE SOUTH SIDE

We've only spoken once briefly on the phone, but Esperanza Guitierrez invites me into her home. I compliment her on the wild and colorful display of perennial flowers that occupies more than half of her front yard and extends across the sidewalk to the edge of Sixteenth Street. "That's my rain garden," she tells me proudly. Although Guitierrez has a tidy and tastefully decorated home, it is clear that she is more oriented toward the outdoors. COVID was an isolating experience; her two kids and eight grandchildren all live in the area, but she tells me that she was unable to see them for a long period of time. "I have a hard time keeping track of time anymore," she says, a condition I can relate to after two years of pandemic-warped schedules.

With a smile on her face, she talks about the Kinnickinnic River Neighbors in Action, a group she has been part of for the last ten years. They held a picnic at the pavilion next to the stream last June that brought seventy people out from the community. The attendees ranged from a six-month-old baby to a ninety-eight-year-old man. She chronicles these events on the Facebook page she created and regularly updates. "The community needs these gatherings," she says. Even though many people were cooped up with the same few people, it is better than being alone. "We all have mental challenges after the pandemic," she tells me.

Guitierrez would know. As a social services worker for the county, she screened calls from people who contacted the Milwaukee County of Public Health, and she listened and attempted to assist those who had numerous challenges. The job had its frustrations, but the steady income allowed her to make payments on her solid brick house in this south side neighborhood of Milwaukee. The south side is the most densely populated neighborhood in Milwaukee, and it is where many immigrants first arrive. Some of the hardest working people who originally staffed the foundries and the tanneries, along with a range of difficult service jobs, live here. Guitierrez seems slightly uncomfortable with the amount of time retirement has afforded her. "I'm sometimes bored out of my mind," she admits. But restoration of the nearby Kinnickinnic River has given her something with which to engage. The Facebook page contains numerous photos of the stream at flood stage, the stream after a winter snowstorm, and videos of ducks paddling up the concrete section undeterred. "I didn't have a voice with the county. Now I have a voice."

Guitierrez's initial involvement wasn't out of love for the river. She had heard that MMSD might be buying homes in her area and was worried that they might want to buy hers. "I couldn't have moved. I was making payments on the mortgage. I was done moving." But she quickly realized that her home, located halfway up the hill south of the stream, was not one of the homes MMSD wished to buy.

Down at the stream, we walk along the trapezoidal channel that confines the Kinnickinnic River. This section is slated to be restored in the next few years (figure 8.1). In preparation for the eventual restoration, MMSD purchased all of the homes on this six-block section that backs up to the stream. The basements of these homes regularly flooded whenever the stream was high and caused the sewer pipes to back up. Heating systems were damaged by these floods. "It was like an open sewer," Guitierrez tells me. Once the homes had been purchased, some of the challenges of the neighborhood crept into the now abandoned houses. Some homes were quickly occupied by drug dealers and others by homeless residents. MMSD agreed to remove the homes and driveways well in advance

8.1 Esperanza Guitierrez is standing next to a remaining paved section of the Kinnickinnic River in Milwaukee.

of the stream restoration and turn the area into temporary parkland, as long as the neighborhood knew that this was a temporary arrangement. The entire area would eventually be lowered, freed from its concrete channel, and set on a new, more sinuous path. Based on the large trees, the vegetable garden beds, the bird-houses, and the public art, it is easy see how neighbors might be content with this interim condition. Guitierrez is impatient to see this section restored; "I tell MMSD that I might not be alive to see this section finished!"

Guitierrez obliquely acknowledges the tagging that has marred the otherwise beautiful displays that honor native cultures: there are still some "jerks who don't respect the park." When she speaks of the gangs, she is far from being a fright-ened grandmother. There is no fear in her voice, only calm determination. "I tell people that we just need to continue to use the park so these bad elements can't get too comfortable." Like the graffiti that will be washed off by a fellow neigh-borhood leader, thanks to a special coating that covers the displays, this is just a speedbump. Every neighborhood has its challenges, she seems to be saying.

Blocks of color have been painted on the concrete channel, testifying to anticipation of the restored condition not far in the future. "Aqua es vida" is written in cursive across the painted mural. Her grandson helped paint this, Guitierrez tells me. Closer to the street she points out the site of a future "water mark," a flashing indicator controlled by MMSD that will light up during flood-ing. The beacon notifies residents that MMSD's combined sewer system cannot fully handle the flow it is receiving. In an ideal world, it might urge residents to delay running their washing machines and dishwashers until the storm is over, and in so doing limit sewer overflows. It will also alert people of dangerous floodwaters, which is particularly important. Water is life, but this river has also brought death at regular intervals. This past spring a child followed a soc-cer ball into the river during a storm; the child's father and a witness followed, attempting to save the child; all three were swept into a section of the river piped underground and drowned. Guitierrez tells me that a version of this hap-pens almost every year.

Guitierrez is a practiced tour guide; she has been part of many tours of the stream over the past several years. She's seen Kevin Shafer giving tours to peo-ple, and on one tour he came over and greeted her warmly. Although she hadn't spent any significant time talking to Shafer before, MMSD and their partners had briefed him on some of the key local leaders. Guitierrez hasn't quite adjusted to the notion that she was the subject of a briefing package, and she mentions that people from England and Paris have come here, taking notes on MMSD's work. Although the attention seemed odd at first, by now she recognizes that something noteworthy is happening in her neighborhood.

We head across Sixteenth Street to walk through the restored section of the stream. Motioning to the ten-foot-tall oaks planted halfway up the banks, Guitierrez says they now have a variety of trees. We cross a handsome pedestrian bridge installed as part of the project, and she says, "I really love the bridge," but the arcane and nonsensical chalk writings on the sidewalk leading up to the bridge unnerve her. She tells me that there is a satanist in the neighborhood who purges himself by writing these unholy missives. Yet on this sunny day in late summer, I can ignore these inscrutable temporary rantings. I look down at the section of the river close to the bridge under the street. MMSD has installed large stones for easy access to the stream as well as comfortable seating. "People come here to just sit and watch the stream," Guitierrez tells me.

From the bridge we look upstream toward the restored section of the stream. She points out the shimmering reflection of sunlight reflecting off the stream onto another bridge. "I love the sounds, the aromas, the sights of this place," she tells me. Guitierrez regularly sees frogs now, but "I never saw frogs before." Wild turkeys make frequent appearances, and one of the most iconic of Wisconsin birds, the sandhill crane, stopped by not long ago. I ask if the social importance of the environment resonates with her. "It always has," she says. "In my culture we say we need to go have a sun bath," and shakes out her arms. "We need it, which means we need to take care of it."

After wandering upstream for about fifteen minutes, we turn around and survey the stream coursing through the re-created floodplain (figure 8.2). On the horizon, the steeple of the Basilica of St. Josaphat is just visible, testifying to the Polish heritage and, despite the presence of the occasional satanist, the strong religious connection of this neighborhood. Guitierrez speaks of the tall grasses on the floodplain, motioning with her hands to show how the water spills out onto the newly created floodplain. She wants me to understand that this stream still floods, but now the water rejuvenates the new plant community. Today the stream is low, and the water that flows below us is particularly clear, a good sign for both the stream and Lake Michigan, which is just two miles downstream from here. We look northeast past the flagpoles that occupy the high ground of Pulaski Park toward Milwaukee Harbor where three rivers meet before emptying into Lake Michigan. As if her thoughts are drifting down with the current, she

8.2 This recently restored section of the Kinnickinnic River runs through Pulaski Park, immediately upstream of Sixteenth Street.

tells me, "Lake Michigan is my cathedral." It's another shared sentiment between us, newly discovered neighbors in this city on a Great Lake.

REBUILDING TRUST IN INSTITUTIONS: A SIMPLE ROADMAP

I reflect on the nature of leadership, a topic of obsession in our culture that celebrates visionaries. Which managerial technique inspires staff to undertake a rapidly expanding suite of concerns and associated challenges? Why is one person able to lead the charge when another cannot? Is it necessary to harvest straight talkers with unassuming backgrounds fresh from the fertile plains of Illinois?

We expect big visions to come from start-up soothsayers and nonprofit leaders. The vision rising from MMSD is born from a government agency, the sector in which public confidence is supposedly in free-fall. In a sector where authority and associated decision-making is spread out among many levels of staff, even small decisions can be delayed because of old beefs between employees, the proverbial sand in the bureaucratic cogs. New initiatives can easily get bogged down, and people can become disempowered.

As I talk to Shafer, I hear something completely different. In his descriptions of various initiatives, I never hear him forwarding a particular solution. His "big hairy audacious goals," as he and his staff like to call them, are pushed on to his staff for a solution, and they are given a long rope. Implicit in this is a recognition that the guy at the top doesn't have all the answers. It also reminds staff that they will have to figure it out. Dave Fowler, a longtime MMSD employee, put it like this: "Kevin does not reinvent the wheel. When we realized we needed a hazardous waste program, he would send us to Duluth in January to see how they did it. When we needed a levee program, he sent us to Tulsa. And when we were starting with green infrastructure, he sent some of us to Seattle."

This expansion of considerations and responsibilities is not standard operating procedure for most government agencies. The approach of most state and federal agencies is tactical retrenchment, whereas the work of MMSD seems boundless. Creating wetlands, redesigning parks, preserving low-lying land in the headwaters, and now paying residents to capture rainwater in their yards, MMSD is tackling the nearly infinite number of ways the watershed must be restored. It represents mission creep in its best sense, which is required to solve the problems that beset our urban streams. With all of these new programs and new responsibilities, one might assume that there was internal pushback. Shafer says that some staff were open to it right away, and others resisted. He simply told his staff that this is how we are going to do it. "Nature didn't screw this up, we did."

I ask Shafer if it is difficult for an engineer to remain uninvolved in the details. He clarifies that "some people call me a closet engineer. I won't question the size of a pipe, but I might question what source of energy is moving the water." I imagine old-line engineers wringing their hands as they realize that their calculations can't provide this answer. For a profession that sweats the details, Shafer continues to question the underlying assumptions.

Even though Shafer didn't provide the answers, he projected an image of a problem solver. Fowler recalls a meeting that followed a series of floods in the 1990s. A particularly conservative political figure who had significant influence on the agency was in attendance. MMSD had a preferred solution that Fowler was prepared to present, but Shafer encouraged him to shelve it for the moment.

In a moment of "verbal jujitsu," Shafer responded to the conservative legislator with a plea for assistance: "We need your help with this." Not pushing a particular solution on the public was the approach that was needed, Fowler recalls, and the long-running cold war with many municipal government leaders and staff began to thaw.

It's dangerous to generalize about a staff of 230 people, but when I replay my interactions with more than thirty MMSD staff over the past six years, I can't remember one person who complained about the work they were undertaking. This strikes me as remarkable. The most subversive response I remember was a comment about Shafer being a "big idea" guy, a frequent gripe from people whose job it is to figure out how to meet these goals. Most staff seem to enjoy figuring out how to meet these goals. Soon after Shafer committed MMSD to producing all of its own energy by 2035, staff members began calculating how they might do that. The first component of this carbon neutrality goal manifested itself in a nineteen-mile pipeline that sends methane from a major landfill to the Jones Island treatment plant. In 2022, that plant is 22 percent of the way toward meeting that 2035 goal. The cost-savings realized by using free methane rather than natural gas makes future investments an easier sell for Shafer.

I suspect it wasn't always easy and ask Shafer where he got bogged down. He says 2009 was an inflection point: "We had been getting beaten up for ten years, and I started to wonder if I could keep doing this. The municipalities contested the costs associated with his proposals, even if the standard alternative involved significant future costs as well." And like everyone in government, Shafer had to answer to someone as well. The board, made up of people representing the range of politics in the region, was generally supportive but not completely on board.

Personality played a big part in his success. My knowledge of Shafer came first from people I met who provided various iterations of "Kevin is a great guy." These testimonies were unprompted, as if the people needed to make the point to anyone who might not know this yet. These comments came from employees and community members. It seemed like they wanted to let me in on a secret—a potential future that offers hope. These people were throwing whatever influence they had behind the guy with the big ideas.

A key moment came in Shafer's 2025 executive director's vision statement, which contained ambitious goals that were light on specific solutions. These documents include promises that can come back to haunt the writer. They go out to anyone and everyone—staff, consultants, politicians, external partners—who has any kind of involvement in the agency's work. This vision statement had the goal of zero CSOs along with myriad programs that might just realize that vision. Shafer admits that he wasn't certain they would actually get there, but "I'm not going to have a goal that shoots for two overflow events." He sent out this vision statement without prior approval of the board, and then he waited for a reaction.

Fowler remembers this email. The vision in broad strokes was already in circulation among some staff, he says, but putting it in writing forced the hand of those who might fund it. It allowed these big ideas to sink in. After a couple of follow-up board meetings, Shafer was not fired and most of his vision was eventually adopted as formal MMSD policy. It took some discussion, but once again Shafer was able to marshal millions of dollars and focus the effort of hundreds of employees. Rather than objecting point by point, the board seemed to realize that they were also going to have to figure it out along the way.

Ambitious goals often seem more fitting coming from a Scandinavian country than from a Milwaukee sewer and water agency. The brazen approach doesn't match the humble package of Shafer the person. I ask him about his special sauce. "When you are in a position like mine, you learn that you have to trust staff. When I came to the ED role, I realized I had to delegate. You delegate to people you trust, and so you hire people you trust." In the end, that may be the simplest definition of leadership—namely, cultivating a network of people willing to carry out a vision, even if that vision is incredibly challenging. Shafer reflected, "maybe I'm not a typical engineer. The people doing this work aren't typical engineers."

CHAPTER 9

COMMUNITY BONDS

Organization and Collaboration in West Atlanta

YOMI NOIBI'S STREAM WALK ON PROCTOR CREEK

We meet behind the regional rail transit system's Bankhead Station in west Atlanta at the trailhead of the Proctor Creek Greenway, an impressively manicured path that runs along the creek for three miles. Beyond the grassy edges of the path, the land is fenced off. Unused warehouses and crumbling parking lots suggest a previous life as a municipal public works lot. A heavy dew covers everything thanks to evening temperatures approaching freezing, a nearly unanimous point of concern in Atlanta. Within the active stream channel, the geological underpinnings of the stream can be traced to the Appalachian Mountains that extend north from here. The water, low after weeks without rain, trickles over granite shoals. The sound of a small waterfall can be heard but not seen because of the dense vegetation. A diverse mix of temperate forest trees grow along the stream; sweetgum, sycamore, red maple, and willows find purchase for their roots in the thin soil above the bedrock.

This gathering was organized by members of Eco-Action at the behest of the Army Corps of Engineers. The Corps has contracted with Eco-Action to engage the community in preparation for a future stream-restoration project on the creek. The Corps has identified eleven locations that are suitable for stream bank stabilization and floodplain creation. Eco-Action, a nonprofit group based in Atlanta, has assembled a mix of people from the area to walk the stream. Their flier asks: "Is it a fish or a tadpole? Only the creek knows—Learn from the creek." Twenty local residents have shown up this morning, and organizers hope to build community support for the property acquisition necessary to realize the stream-restoration project. Stream restoration is as much about shaping people's perceptions here as it is about shaping the channel.

Organizing the walking tour and meeting today are a group of semiretired men who come to this work from vastly different stations. Yomi Noibi is the former director of Eco-Action and a former professor at Spellman College who is attempting to slow down his work after thirty years of community organizing in Georgia. A soft spoken seventy-year-old with a warm thick accent from his native Nigeria, Noibi is known by everyone. He has assisted with most of the community engagement work in this watershed in some form or fashion. In addition to numerous EPA Region 6 funded efforts, Noibi has assisted with the Atlanta Watershed Learning Network effort, a program aimed at providing watershed training to both adults and youth. Most of the people I will meet in the coming days have attended this training or a similar one that Noibi has helped organize. After brief introductions, Noibi implores everyone to talk to one person who they have not spoken to yet. "This isn't just about stream restoration," he says, "this is part of the process."

The other organizers of today's tour are Keith Parsons and Bill Eisenhauer. Parsons is a retired biologist with the state's natural resources agency, and in his former role he reviewed numerous stream-restoration plans for technical and regulatory compliance. He volunteers his time with Eco-Action and the Georgia Rivers Network when he is not fishing. Eisenhauer, a northerner who established himself in Atlanta after gaining a degree in chemical engineering from Georgia State University, is a self-described "behind-the-scenes guy." Having done very well in local real estate investments, Eisenhauer volunteers his time for seemingly every environmentally related initiative in the metro region. He seems to know everyone, and everyone knows him.

Although the maples have turned a beautiful shade of crimson, Proctor Creek is not showing its true colors at the moment. The defining element of this creek is its propensity to flood as it winds through a multiplicity of challenges that range from economic to ecological and everything in between. The uppermost reaches of the watershed are dominated by large, institutional properties. A good portion of the runoff from the Atlanta University Center, a consortium of four historically Black colleges and universities, drains into Proctor Creek. The new Atlanta Falcons stadium and its parking lots drain into it as well. Just downstream from these properties lie neighborhoods that struggle with some of the highest poverty and property vacancy rates in the city. If that wasn't enough, the regular floods punish those who live here and highlight the poor decisions of those who constructed housing in the low-lying areas of the watershed. Despite these challenges, the area has seen massive investments. Drawing on the financial heft of the U.S. Army Corps of Engineers would be the ultimate goal, but the Corps does not have a history of funding these smaller, more costly urban stream projects.

We walk down the path past mile markers and signage describing this as a "Remote Area." This section is only three miles long, but it connects to an ever-increasing network of paved trails that feed into the Atlanta Beltline, a major effort to build a nonmotorized path that will encircle the city. Today our group scoots to the edges of the path on several occasions to accommodate approaching bicyclists. We stop at a bridge that overlooks a rocky section of the stream. Parsons brings up the issue of how the stream might have looked two hundred years ago and how that might inform the stream-restoration designs. The group doesn't seem to grasp the relevance of this because the context is so different now (figure 9.1). Parson begins to explain that the stream, with its

9.1 A building that occupies space on the floodplain of Proctor Creek in west Atlanta.

bedrock base, still retains the structural integrity that can lead to a successful restoration project. The challenge is the lack of floodplain storage.

One attendee dressed in a camo jacket, black pants, and a black knitted cap shakes his head. After Parsons's discussion of historic conditions, he asks if anyone has gotten flooded. "Let's be relevant," he says bluntly. Perhaps hoping to provide an alternative target, Parsons brings up the looming development plan that most everyone appears to know about. Most of the land surrounding the path we have just walked has recently been acquired by Microsoft. The ninety-two acres will become Microsoft's southeast headquarters. Parsons explains that the community needs to demand world-class stormwater management. He suggests that the residents ask developers and the city what they are providing in terms of floodplain storage. Parsons repeats the term *floodplain storage* in anticipation of some future battle. No one seems to know what is going on with the Microsoft development or how the community might be able to weigh in on it. Even the ever-connected Eisenhauer says he is still looking for a connection with the company.

This topic increases the camo-jacketed man's irritation. In response to a prompt by Parsons for thoughts from the group, the man says, "It's a land grab, that's what it is. Just look at the natives that used to live here. That's what's going to happen to us." The man says he's going to leave, apparently not satisfied that this discussion of the stream will rectify any of these broader concerns. In sequence, several people sound off and ask him to stay. Noibi heads over to the edge of the group where the man is stewing. Whatever Noibi says is unclear, but between his thick accent and intense attention, the man is calmed and remains with the group. Noibi rests his hand on the man's back. The interaction has a deeper, undefined quality; fatherly, spiritual, or perhaps just communal. In fact, Noibi has rested his hands on several people's backs in the past hour in a manner that seems both foreign and comforting after the isolation of the pandemic. In a culture that often devalues age and wisdom, Noibi is showing a youth-obsessed America the benefits of listening to their elders.

Noibi addresses the burden of history that is presently marring an otherwise beautiful morning. This is what Eco-Action has been working on, he says (figure 9.2). "We may not be able to go back to historic conditions, but we can go to a healthy watershed. We have to correct history. We know of these terrible things—do we want to continue that? No, we can do better." The specifics of the path forward aren't clear, but the intent is. The assembled group seems to

9.2 Yomi Noibi is leading the discussion about the future of Proctor Creek surrounded by a group of interested residents.

understand the broader goal to which Noibi is referring and is happy to let his optimism seep into their consciousness.

CHANGING ECONOMIC REALITIES

We cross the busy highway and walk down a former residential street toward the stream. After construction of the regional rail transit station in the 1960s, the flooding the neighborhood already experienced was exacerbated. People were relocated, and homes were demolished even though the land was not fully purchased by the city. Nothing seems to have happened here in the past sixty years, and the site has turned into an urban wilderness. In the absence of human activity, trees have grown to impressive heights and are somehow surviving the onslaught of kudzu, which creeps over anything less than twenty-five feet tall. Chris Theal, codirector of the Proctor Creek Stewardship Council, has mowed paths through the kudzu with his tractor. Without these paths, several of the older attendees would have trouble navigating the site.

This location is known as PC-21, one of the areas selected by the Corps of Engineers and the city for active stream restoration along Proctor Creek. City worker Julie Owens passes out handouts with project descriptions and maps so people can locate their position. The sheet she has circulated calls for construction to begin in January of 2021, wrapping up in December of 2024. Eisenhauer mentions that this puts us about twenty-one months behind schedule. Owens, an engineer for the city's Watershed Management Department, is the point of contact person for the project with the Corps. For the project to move forward, the city must secure ownership of all of the private landholdings. She begins her explanation of the current status by talking about how the city offered to buy three lots from one woman. The city offered $15,000 and the homeowner came back asking for the assessed value of $20,000. The city agreed, but Owens stated that, unfortunately, they didn't seal the deal at that moment. As soon as Microsoft purchased the property across the highway, the assessed value jumped to $150,0000, even though the land is in the floodplain and is unbuildable. The city's envisioned contribution to this partnership with the Corps that encompasses eleven restoration sites is the acquisition of a hundred parcels. They have a real estate budget of $700,000.

Owens shrugs and mentions that "we'll see what the Corps does," realizing that these new economic realities might change everything. The city has already spent $500,000 on just three parcels at this site. The estimated $8.6 million budget for all of the stream-restoration sites on Proctor Creek could be expended on this one site alone, she tells me. "The federal process is even longer than the city process," she bemoans. The Corps (not present) serves as a convenient scapegoat. Given the city's delay in purchasing the properties, it's not clear who is dragging out the project. Without any feds being present, there is no one to argue with, and the project's status appears uncertain.

When we get to the creek, the group edges toward the precipice of the bank, some fifteen feet above the water. Kudzu covers the edge of the bank and creeps up smaller trees, creating a ghostly and penumbral blanket of undulating vegetation. This ubiquitous vegetation of the new South conceals the soil below; tree roots are still intact but the leaves are struggling for sunlight. On a practical level, the kudzu obscures the edge of the bank. I'm not sure how far out I can stand without falling down the bank. I put my trust in the stability of its root network;

it's clearly thriving here. The stream is braided and wide—too wide, Parson says. He talks about how the Corps will likely use the soil below us to create a floodplain where we now stand with an active channel on the far side. Sturdy stone control structures will be installed to ensure that the stream doesn't chew into the newly formed floodplain. These structures will be needed because this area drains some 1,652 acres of highly impervious land.[1] Some kids across the stream shriek in a call reminiscent of a red-tailed hawk. Chris Theal engages them in discussion. They are skating in a parking lot that we can't quite see through the trees. At a stream, unexpected interactions seem to pop up regularly.

Standing on the bank of a stream whose future shape and form rests on murky land purchases and opaque internal operations of the Corps of Engineers, the group appears to be supportive. However, there isn't much for them to do. Some of the younger kids start to chat among themselves. I think about the challenge of engaging residents in a process driven by real estate transactions and dealing with different government agencies. How can one advocate for action on the stream when it isn't clear who is making the calls? When economic forces descend on a neighborhood, it's not clear that anyone has the power to stop them. Despite this insipid uncertainty, the members continue to chat among themselves. At a minimum, as Noibi encouraged at the outset, new connections have been made.

DONNA STEPHENS AND BILL EISENHAUER: ADDRESSING THE PAST, IMAGINING THE FUTURE

The stream project is just one component of the broader efforts going on in this watershed. In the last fifteen years, a flurry of planning has examined and identified opportunities to capture the excess stormwater runoff that plagues both the stream and the people living close to it. One study published in 2010 was facilitated by a local parks organization, Park Pride.[2] Known as the PNA study, this study identified numerous opportunities to capture stormwater by creating new parks that would function both as gathering spaces and stormwater facilities. These parks were proposed in areas that already flooded and oftentimes had a high number of abandoned houses. By purchasing these derelict properties and turning them into attractive greenspace, both the stream and the people living there would benefit. Realizing the broader vision would require millions of dollars for property acquisition and millions more for the construction of the parks.

Since 2010, numerous entities have set out to accomplish this vision. They have made significant strides. The fact that they have so far only captured a portion of the stormwater necessary to eliminate flooding does not detract from their collective accomplishments. The projects range from smaller scale neighborhood parks to larger parks that span multiple city blocks. The methods of stormwater capture range from curbside bioretention to large ponds to underground cisterns.

I ask Eisenhauer and Donna Stephens to show me some of these parks in the English Avenue and Vine City neighborhoods of west Atlanta. They are both very familiar with the neighborhoods and with each other, having engaged in many of the same meetings, trainings, and events over the past decade. We pile into Eisenhauer's car and begin a circuitous tour. When I ask who is coordinating these multiple efforts, Eisenhauer stresses that the work is grassroots in nature. No one entity is leading the charge, but a long list of organizations and foundations assist in different ways. I suspect that the decentralized nature must lead to organizational turf battles or to unwarranted largess for favored projects.

At first glance, I am inclined to level the latter accusation toward the largest of the stormwater parks, Rodney Cook Sr. Park, which includes raised concrete paths that wind gracefully through a space equivalent to five city blocks. Before this investment, this area was another regularly flooded neighborhood that had been bought out by the city. Although the stream is piped in this headwater area, under heavy rains the pipes would overflow and surface runoff flood homes and streets. The area sat empty for more than ten years before the current configuration came to fruition. Gleaming stainless-steel railings line the paths that lead to overlooks where I can see placid, open water that is edged by pickerel weed and rushes. The polished nature of the park brings an undeniable sense of prominence to this place that, by most other indicators, had fallen on hard times. In this sense, the expense seems fitting. It also matches in scale and ambition a similar park located in a wealthier part of town on the east side. The historic Fourth Ward Park, often criticized for the rapid gentrification that it precipitated, nonetheless has turned its own east side neighborhood into a bustling hub reminiscent of New York City's Highline. Few could fault the developers of Rodney Cook Sr. Park for cutting corners or skimping on the fixtures (figure 9.3).

It is in the dual nature of restoring urban watersheds that the vast need is met by vast potential. Every street and parking lot has potential for retrofitting. Theoretically this is a task that has no end, particularly when we factor in

9.3 Donna Stephens and Bill Eisenhauer at Rodney Cook Sr. Park look out over new, elevated walkways and aquatic beds designed to hold stormwater runoff from the neighborhood.

more frequent and heavier rain events anticipated with climate change. It is the ultimate challenge that seems to lodge in certain people who have grand ambitions. As we drive through the English Avenue and Vine City neighborhoods, Eisenhauer points out opportunities ranging from street retrofits to stream daylighting opportunities. It's hard for an outsider to discern the viability of the projects, which depend not only on technical considerations but also on landowner buy-in and funding. I'm not sure to what degree Eisenhauer's Herculean list of projects is based on hope or reality. With no formal role, Eisenhauer speaks of getting funds for this or that project as he mentions past meetings with the mayor or local commissioners. He projects the demeanor of a jocular, old-time party boss with his hands on the levers of power. Originally from the north, Eisenhauer has deftly adopted the soft power of the South that is lubricated by relationships and connections. It appears that the influence he wields is aimed at the goal of restoration of an entire watershed: the water, the land, and the people therein. The immense scale of this endeavor seems to meet his ambition, intellect, and activity level. Eisenhauer has his eye on the prize.

Stephens has become a key member in west side environmental work in the past decade. Her deep and deliberate voice encourages a slower and more thoughtful discussion. She became involved first by participating in the inaugural class of the Atlanta Watershed Learning Network (AWLN) organized by Noibi. At the same time, she took part in an Environmental Justice Academy organized by the EPA. She told me that she had been attending two meetings a week for about six months. It was an education that paid dividends. "I really learned what green infrastructure meant," Stephens said. A watershed moment was participating in a field trip to the Mercedes-Benz stadium, $1.8 billion home for the Atlanta Falcons and the Atlanta United FC. This shining testament to the wealth generated by professional sports sat at the top of the Proctor Creek watershed. Through the efforts of many, the new stadium captured significant volumes of stormwater. It was a lesson to see how development could be harvested to benefit the stream and provide economic benefits to the community at the same time.

Since her graduation from these classes, Stephens has been appointed to the boards of Groundwork Atlanta and Historic Atlanta, has cochaired the Proctor Creek Stewardship Council, and founded the Chattahoochee Brick Company Descendants Coalition. The latter was precipitated after "nerding out with her sister on PBS." They came across a documentary based on Douglas Blackmon's book *Slavery by Another Name*. In the book and the documentary, Blackmon tells the story of convict labor that began just three years after the end of the Civil War and continued until 1908. Stephens watched the show and realized that the site of the former Chattahoochee Brick Company, which inflicted terror on thousands of mostly Black men, was less than a mile from her house.

The brick company is located where Proctor Creek empties into the Chattahoochee River. Its rubble remains embodies the complexity of restoration in both the ecological and the social sense. The soil requires clean up, and properly recognizing this injustice will require coming to terms with the legacy of convict leasing. But the land can also connect with broader efforts aimed at preserving land along the Chattahoochee River, and it may offer new greenspace to neighboring communities.

Stephens speaks about her work on this project with a gentle conviction. Although she may be surprised by this turn of events in her life, she doesn't tread lightly in her efforts. She recently emerged from a contentious battle with the Norfolk Southern Railway; they wanted to turn the former brick site into a transfer station. With help from the Trust for Public Land, the parcel located at the confluence of Proctor Creek and the much larger Chattahoochee River has been purchased. Remediation of the soil is currently being analyzed. The future use of the site hasn't been decided, not to mention how to properly acknowledge this shameful part of America's history. Difficult conversations are on the horizon. In a recent panel discussion led by Blackmon, Stephens said "the need for greenspace is monumental." But she speaks to the broader challenges for the people living in this area, citing her niece who has to commute almost four hours each day to her new job. "The door is constantly closing on a certain population that is primarily Black and low income."

As we head down toward the former brick company site, Eisenhauer peppers Stephens for navigation tips. Eisenhauer, two steps ahead of everyone, gets testy with Stephens, and she returns the favor when he misses the next turn. The traffic is thickening in the late afternoon, and we still have to cross town again. Eisenhauer breaks the ice, saying that "we must sound like an old married couple." I smile and tell them, "well, I wasn't going to say it." Much like with the traffic they are fighting, these two seem intent on find a path through the legacy issues that haunt this part of Atlanta. This odd couple shares a passion for making things happen.

ANNIE MOORE: HYPER-LOCAL STORMWATER MANAGEMENT

I was first introduced to Annie Moore by Noibi when the two of them were in my hometown attending a conference. She had taken the AWLN class, and through her work on the park had been invited to be part of the Atlanta delegation to the One Water Summit. She invited me to take a look at her work and the neighborhood. We meet on an unseasonably cold morning at Lindsay Street Park. Although abutting the park, there is no official border between Moore's home and the park. Her home and the adjoining park stand as an oasis in a social and economic storm. Even in late October, the late blooming flowers tower above the street. Moore has become the lead caretaker of this park and heads a small work crew that maintains three of the new parks in this area. She explains that she and her team don't use any herbicides in the park. She leads me to a mass of yellow sunflowers that extend a couple of feet above our heads and frame the

9.4 Annie Moore standing next to her plantings at Lindsay Street Park.

park sign. "They're supposed to grow to only six feet, but I guess they like me," she says with a gardener's pride (figure 9.4).

Moore's yard backs up to a tributary of Proctor Creek that runs above ground for a couple of hundred yards before reentering a storm pipe. The creek winds through the park under the shade of massive overcup oak trees whose longevity point to the history of the neighborhood. Initially established by a predominantly white working class, it was abandoned during the white and middle-class Black flight that spanned the 1940s through the 1960s. The history of flooding has affected all who have lived here throughout the years. That historical arc snaps back to the present when Moore shows me where recent torrents of stormwater have jumped the curb and flowed toward her foundation.

This park was built with the help of the Conservation Fund and was the first one with a goal of reaching underserved neighborhoods. The Conservation Fund has also contributed to the maintenance by supporting Moore and the seasonal green team. Team members earn between $15 and $20 an hour for maintaining the green infrastructure components of the park and other plant beds. The city

Parks Department continues to remove the trash and mow the lawns. Moore frequently mentions her run-ins with park staff when they ask what she's doing with this or that plant bed. She always tells them to leave it alone. They usually listen to her.

The park contains tidy stormwater planters and artfully curved channels that connect to a large rain garden. There is a small but attractive playground and a few interpretive signs that describe the purpose of park elements. The beds are mostly turning to the duller shades of winter, and Moore leaves the stems there over the winter as cover for pollinators. She is most animated when speaking of the plants, engaging in this aspect over which she has full control. As we take it in, an older woman approaches and asks for money. Moore gently mentions that she doesn't have any but that the church up the street is providing breakfast this morning.

We walk further into the park where a stand of large trees covers the creek. Without fanfare she alerts me to watch out for the used needles in the path. We hop over the creek, and Moore explains her plans to create an access point on this side of the park. Where water currently flows over the turf and has eroded the soil, she tells me that a modified rain garden will be installed to slow the runoff. Park Pride has already earmarked $75,000 for the project, and Moore hopes that another $75,000 will be available to fully realize the project. In its current state, it's difficult to imagine this creek causing so many problems. The water, only a couple of inches deep, wanders around the base of stately trees. Standing on the grassy bank, I can imagine the baptisms that long-term residents describe as having occurred in the creek. More a source of fear than redemption today, no one has used the stream in this ceremonial manner in several decades.

We head down to Kathryn Johnston Memorial Park, a larger park only a few blocks away. Recently remodeled homes sit next to homes boarded up and fenced to keep squatters out. The blinking blue light of police cameras mounted on the light poles at each intersection signal that crime is still a challenge here. Along the way, Moore points out a recently remodeled home where the owners poured a concrete driveway and walkways that extend to the house, sealing the property in an impervious bubble. She attempted to help address the counterproductive paving and stormwater it would generate as part of her capstone project for the watershed class.

Just down from the park several newly built two-story homes with front porches have been alternately painted in muted shades of blue, yellow, and green.

Their tidy, boxy frames stand ready for new homeowners who can afford the $285,000 asking price. When I ask about concerns over gentrification, Moore doesn't seem concerned. "There are too many Black people over here," she says frankly. But the perceptions of a long-term resident don't always match the eternally ambitious perspective of real estate investors. The real estate listing descriptions for the area claim "Incredible opportunity in HOT English Avenue area," and "unbeatable location 10 minutes from Downtown, Georgia State." In a city with a severe housing crisis that is experiencing an influx of residents from more expensive real estate markets, every patch of ground has value. Although she doesn't like the new homes, Moore appreciates the rehabilitation of several apartment buildings next to her house. The local church purchased and remodeled these four- and six-unit dwellings, and many of the people who once lived there have returned.

Despite Moore's enthusiasm for the plants, she seems skeptical about any broader benefits for the neighborhood. Long-term funding for the green team is uncertain, and although appreciated, it doesn't amount to a life-changing sum. It seems that Moore will care for this new park regardless of who moves in and whether or not she is paid to do it. She tells me of her plans for terraced beds between her house and the park. Facing this area and looking over the park, there are five large, colorful paintings of otters, frogs, and cranes. Collectively they shout a message of biophilic appreciation to park goers and the pollinators that frequent the gardens. They serve as a clear welcome sign to nature that is being slowly cultivated in this park. Under the shade of a particularly large overcup oak tree that anchors her corner, we stand next to a small fig tree that Moore planted. It is graced with a few unripe figs. She tells me that it isn't thriving next to the curb where it has been planted. She's considering moving it closer to the stream.

ANOTHER NOIBI DISCIPLE

Another member of this loose network of community activists is Al Tucker, who saw a flier advertising the AWLN training that promised "come walk Proctor Creek." This piqued Tucker's interest because of his memories of exploring the creek as a young kid. Tucker moved to the Hunter Hills neighborhood in 1949 when "everyone in the neighborhood had an outhouse." In addition to the occasional neighborhood mule, most folks had chickens. "People moved to Atlanta from the country and kept their country ways," he explains. Tucker's watershed education eventually came in 2020 when he took the class remotely because of the pandemic. At first he didn't realize that the purpose was greater than walking and learning about the creek. He soon learned about "the other 90 percent; the environmental justice component."

As Tucker explains Atlanta's advisory structure of Neighborhood Planning Units, I realize that watershed training has understandably morphed into political advocacy. Tucker is eager to relate the twists and turns of past and present hot button issues where residents have weighed in on them. There have been some successes and some failures, he explains. He counts Stephens's work with the former Chattahoochee Brick Company as a major success. "The training shows how to influence authority," he says.

"Dr. Yomi has given me this 'ambassador' title," he explains, somewhat self-consciously. "Dr. Yomi, you know, he's a bottom-line person. He can connect the dots." "Everywhere I go, people already know him." Tucker is spending his retirement producing the newsletter, helping with recruitment, and "helping people stay engaged in the training." He plans to keep at it, he tells me, "as long as there is breath in my body."

COMMUNITY ORGANIZING AND PERSONAL CONNECTIONS

About a month earlier, I had met with Yomi Noibi in Milwaukee. He had heard about the Urban Ecology Center and wanted to see it. I was happy to go with him and, in the process, try to get a better handle on what this community organizing work entails. We enter this well-known nature center, and within minutes Noibi is engaging with the two staff members at the front desk. He tells them that he knows about their work and wants to bring some of this to Atlanta. He is expanding his connections. We walk outside and look out toward the river. He would like to get up to Green Bay to reconnect with a student from three decades ago. This student, a Menomonee tribal member, had explained the ways his community organized when he was in Noibi's class.

Later he shares that he had a challenging discussion today during the conference. During a session focused on inequity and environmental racism, a colleague from the Watershed Department of Atlanta was less forthcoming than she had been in the past. "There is a new boss in town, and things have changed.

I understand." But the loss of connection between this person he had seen as a member of his larger orbit was bothering him. "When you break bread with a person, you are family," he tries to explain.

I ask about Noibi's first experience with community organizing, and he described his first task with Eco-Action thirty years ago. His new boss had sent him to a completely white rural Georgia county that had been trying to address issues with illegal dumping of tires and other trash. Participants' initial response to him was neither positive nor negative, only puzzled. But he worked the process he was taught: "Assess the needs and problems. List the problems. Prioritize the needs," he recounts. After a couple of months, the community had made some progress on their issue. "They were happy!" They were telling me that "that was easy." He laughs, "I reminded them that it's not necessarily easy."

CHAPTER 10

SOUTH RIVER ACTION HERO

Environmental Justice and Activism in Suburban Atlanta

DIFFERENT WATERSHED, DIFFERENT RULES

I'm riding with Jackie Echols, board president of the South River Watershed Alliance, on a road near the headwaters of South River in suburban Atlanta. The river drains the urban and suburban portions of DeKalb, Cobb, and Clayton counties and winds sixty miles in a southeasterly direction before emptying into Jackson Lake, a 5,000-acre reservoir created when the Lloyd Shoals Dam was constructed in 1910. At this point, it combines with the Yellow and Alcovy rivers to form the Ocmulgee River, which winds through the lowlands of Georgia before feeding into the Altamaha River, Georgia's largest river.

We're less than eight miles away from Proctor Creek on the west side of Atlanta, but this river drains into the Atlantic Ocean rather than into the Gulf of Mexico, one thousand miles distant. The dividing line for the Atlantic drainage basin runs right through metropolitan Atlanta. Less evident in an altered urban environment, the boundary still carries significance. It highlights the importance of preexisting terrain in defining vastly different paths. Depending on the vagaries of a storm gust, a water drop falling on that dividing line might end up in the bayous of Louisiana or on the sandy beaches of the Georgia coast.

In Echols's Subaru, we drive through a landscape populated with a mix of mundane but generally undesirable properties: juvenile detention facilities, asphalt milling and storage lots, illegal dumps, and warehouses. Nearly all of the commercial properties are fenced off, and most have razor wire to make the message clear. Echols tells me, "No one cares about this place or the river." Due in no small part to Echols, this frustrated sentiment may be changing. In 2021,

American Rivers listed South River as the fourth most endangered river in the country.

Echols turns off a trunk highway and goes down the road, passing houses that were probably surrounded by farmland when they were built. Today their neighbors are metal fabricators, auto wreckers, and a road construction contractor who stores a forty-foot-tall pile of asphalt next to the road. The road dead-ends at the railroad tracks. On the other side of the tracks sits an illegal dump with piles of debris extending skyward in a near vertical fashion. Echols heard about this unpermitted dumping ten years ago when they were doing clean-up at nearby Constitution Lake. She notified the county, but "it's one of those things that happens in this area—there was no big effort to stop it."

I have made similar windshield surveys of watersheds in the past. I understand what frustrates Echols. As she drives she points out different development actions that have not gone as she would have hoped. It's clear that cleaning up this river will require decades of effort to address the thousands of small, but cumulatively significant impacts. Removing impervious surface that causes flooding and flushes pollutants directly into the stream requires a significant outlay of funds on marginal properties where few owners care about these issues. Stewards of such a watershed are left playing a very bad hand. Perhaps the best that can be done is to bluff your way into the game and hope to score a couple of wins.

Despite these challenges and a deep well of accumulated frustration, it's evident that Echols cares about this particular area of suburban Atlanta. We begin in the upper watershed in East Atlanta, a predominantly middle-class white neighborhood where she has some business related to an upcoming fund-raiser. After dropping something off at a house, we follow the path of Intrenchment Creek south (figure 10.1). As the river widens and approaches its confluence with South River, the makeup of the population shifts imperceptibly. Echols explains that this trend extends the length of DeKalb County: the northern third, which drains into the Chattahoochee and eventually the Gulf, is predominantly white, and the southern portion that drains into South River is Black.

The racial differences parallel economic differences that are manifested in development patterns. Development in a booming metropolitan Atlanta is not equally distributed. In northern DeKalb, communities such as Dunwoody and

10.1 Jackie Echols, board president of the South River Watershed Alliance, at Intrenchment Creek.

Johns Creek boast schools that receive A and A+ overall grades, respectively, from the Niche school ranking website.[1] Development there includes new homes and commercial entities much more likely to be serving bagels than fabricating metal. This development is a major reason sewer infrastructure investments have been prioritized northern DeKalb, according to Echols. Southern DeKalb is a different story.

In southern DeKalb County, recent development plans follow a familiar pattern of commercial use on larger, secured lands. Two planned developments have doubled-down on size and are targeting some of the last remaining forested tracts in the area. These two adjacent forested tracts sit in the center of a broader 3,500-acre mix of land called the South River Forest. The organizers, of which Echols is one, describe this future South River Forest as "a network of civic amenities united by a lush forest canopy and the headwaters of the South River." In the southwestern corner of DeKalb County, adjacent to the city of Atlanta, many see these forested tracts as a chance to lock up greenspace for a growing population. The Nature Conservancy has described it as "one of the most ambitious concepts for greenspace expansion in Atlanta." From a watershed perspective, this forest can cushion the impacts of the impervious landscape that surrounds them. For the river, these forested tracts are one of its biggest assets. Despite this, they are slated to be chopped up by a mix of players that one might not expect.

The first is a proposed land swap the county arranged with Ryan Millsap, former owner of Blackhall Studios. Atlanta has become an increasingly desirable location for movie and TV production nationally thanks to agreeable weather and generous tax credits provided by the state. In 2020, DeKalb County agreed to swap forty acres of land in Intrenchment Creek Park for a slightly larger parcel of floodplain land. The forty acres in the park are located on the highest point of the 130-acre park, compromising both the recreational use and the ecological integrity of the park. When the swap was made, Echols explains, the forty acres was valued at $3.2 million, which allowed the county to trade the land for a larger floodplain parcel valued at a slightly higher amount. The forty acres is now worth more than $10 million. This current value is significant given that Millsap recently offloaded the production studios and a significant amount of bad press to a private equity firm, the Commonwealth Group. His new endeavor, Blackhall Americana, will explore quintessentially heroic American figures: "Americans are a culture of bad asses," he explains. It's safe to say that Millsap probably had Matt Damon or Steven Segal in mind rather than a five-foot-three-inch Black woman with a PhD in political science. His understanding of bad asses may need to expand beyond beefcake action heroes to include environmental activists.

Not familiar with movie production facilities, I wonder what the impact of such a facility might be. Echols drives me by Blackhall's existing movie production facility, and through the fenced entry point I notice numerous nondescript black structures. "Nothing but asphalt and warehouses," Echols says. New industry, same footprint, similar negative impact on the watershed. This land swap between a production magnate and county officials has produced a significant outcry from the community. In February of 2021, the South River

Forest Coalition and the South River Watershed Alliance sued to stop the swap. The legal complaint states:

> The land exchange represents an unlawful conversion of public park land to private uses and a waste of taxpayer money. The land exchange violates the conditions imposed via deed on Intrenchment Creek Park, which conditions may be enforced, and any member of the general public who utilized the Park. The land exchange is not in accordance with laws and regulations concerning the use and disposal of County property.[2]

"Everything is legal until you sue," Echols says. She expects the judge's ruling sometime in the next six months.

COP CITY, URBAN CANOPY, AND LOCAL POLITICS

Another major development has unleashed a simmering standoff unlike anything else in Atlanta in the last twenty years. In April 2021, former Atlanta mayor Keisha Lance Bottoms and the Atlanta Police Foundation announced its plans to build a $90 million cutting-edge police and firefighter training facility on land owned by the city in neighboring DeKalb County. The 350-acre property, known as the Old Atlanta Prison Farm, borders Intrenchment Creek Park. The plans include a burn tower, a space for high-speed chases, and a mock city that will allow police to practice urban conflict scenarios in a realistic urban environment. This has raised many eyebrows following countless police shootings of innocent Black people and the subsequent Black Lives Matter protests. More than 1,110 Atlanta residents have provided recorded comments on the development, which took two days to play out loud at the council hearing. More than 70 percent of the comments opposed the development.[3] The city council approved it anyway. The city is on the hook for a third of the cost, and the Atlanta Police Foundation, supported by several influential civic leaders, will pick up the tab for the other two-thirds of the cost.

A leaderless group of mostly anonymous protesters has occupied the site with tents and treehouses. They have labeled the site Cop City. A tow truck was recently burned, and security cameras have been destroyed. Media outlets from around the world descended to cover the story. Echols has released press statements on the issue and has spoken to reporters from Al Jazeera, the *New Yorker*, CNN, and *The Daily Show*. Although Echols protests this development, she tells me she's not "anticop." Her focus is more on the lost tree canopy than on the militarization of the police. She sees the value of the police but doesn't see the need for this particular facility.

Although the issue has become a mega-fire that creates its own weather, she tries to steer the focus back toward the watershed and the need to protect these last remaining forested parcels. She, along with a broader coalition, is advocating for establishment of the South River Forest, a roughly "3,500-acre conservation area that would unite and protect the network of natural features in this area."[4] The coalition envisions an ecologically resilient swath of nearly contiguous forest canopy with the addition of a number of new civic amenities. Most of the area is currently open space, but some parcels would need to be purchased. Local environmental organizations have called it one of the most ambitious concepts for greenspace expansion in Atlanta. The current threats are troubling because they take aim at the heart of the proposed conservation area when the concept of protecting this forest is gaining steam.

We pass the lot where the tow truck was burned, and Echols tells me that not long ago a reporter got her car tires slashed. We decide not to trespass or risk entanglements with either security guards or anarchists. The property contains a mix of thirty-year-old forest interspersed with numerous larger trees and remnants of former prison buildings. With the debris cleaned up, management of invasive plants, creation of trails, and adoption of some programming, this site could become a prized recreational destination for all of Atlanta. The city seems to be of two minds about the project. The city's Department of Planning has endorsed the concept of the South River Forest in principle, but the Atlanta Police Foundation has received the support of key political and financial brokers in the city. DeKalb County does not own the property and has a limited capacity to stop the development. DeKalb County commissioner Ted Terry introduced a resolution calling for a full environmental assessment of the plan but "got skittish" when concerns about antagonizing Atlanta city council members arose.

Echols is looking for an angle. Others have occupied the area, building tree forts and enduring run-ins with the police. Echols is trying to get into the land disturbance permit that must be issued by the county. Given the existing total maximum daily load (TMDL) for sediment in Intrenchment Creek, the sediment the 85-acre site could generate would probably violate the TMDL. Issuance of the land disturbance permit does not require a public hearing, but the documents must be available to the public. Another angle is the probable use of fire retardants that contain high amounts of PFAS, commonly termed "forever chemicals." Although the Atlanta Police Foundation maintains that these would be directed toward a sanitary sewer, Echols believes pollutants would escape into Intrenchment Creek.

Echols seems to be playing catch-up, trying to assemble some evidence of the pollutant loads of such a massive land disturbance. She's enlisting the help of a stormwater engineer to gather information and put it in front of regulators whose job is to uphold the TMDL limits. The City of Atlanta and powerful funders are firmly behind Cop City, so the odds seem stacked against her. But Echols isn't ready to throw in the towel or offer a pronouncement. In her world, it is all still in play, and she will continue to work her angles.

We stop at a golf course that stands out as an amenity in this area. Echols needs to collect a donation for an upcoming fund-raiser for the South River Federation. She requested a few rounds of free golf as a potential bid item. The golf course manager told her that he'd need to structure it in a different way because his management company works for the county. Apparently Echols has bruised a few egos in the county bureaucracy. "He's a nice guy. We worked it out," she says as she returns to the car with an envelope containing a gift certificate.

TOOTHLESS CONSENT DECREES: LITTLE CONSENT AND FEWER RESULTS

Behind the scenes transfers of land are only one of the challenges facing the South River watershed. The single major cause of unhealthy water quality are high levels of E. coli bacteria stemming from human waste. The two vectors for bacteria are combined sewer overflows (CSOs) located in Atlanta and leaking sanitary sewers in DeKalb County. Echols has been working on both of these issues for many years. In a pattern unfortunately common to many urban areas, the resolution of long-standing sewage issues drags out for many years. The level of investment and the change required to effectively correct the aging infrastructure is so massive that most bureaucracies are unable to address it in a proactive manner. In most cases, it takes a lawsuit. This precipitates a negotiated consent decree overseen by a local judge. There have been many consent decrees that effectively determined the water quality of South River, and by extension the ability of residents living there to use the river.

The first consent decree related to the City of Atlanta's combined sewers. This consent decree, initiated by a citizen plaintiff in 1995 and eventually signed in 1998, required the City of Atlanta to address the overflows from approximately fifty to seventy per year down to four; $2.5 million in fines were levied against the city.[5] It is important to note that the Department of Justice (DOJ) required that some parts of the combined system be separated. Separating the sewers would limit the number of overflows in the future. The DOJ required that environmental justice concerns be considered when selecting which sewers were to be separated. This consent decree resulted in improvements to South River; approximately nine miles of combined sewer were separated, which reduced the number of overflow points from three to one. Although the consent decree resulted in improvements in South River, Echols states that the river has not consistently met water quality standards for bacteria. Leaking sanitary sewers, most in DeKalb County, continue to plague the river.

These leaking sanitary sewers are the second major source of *E. coli*, and they were not addressed in the first CSO consent decree. This would require a separate consent decree with DeKalb County. Initiated in 2010, this court-ordered mandate was intended to force the county to replace broken and undersized sanitary sewers. Leaks had been noticed for decades, and attempts at repairs were insufficient and sporadic. Echols believed this consent decree was the missing piece in addressing the primary sources of bacteria. Addressing both would put South River on a path for safe recreational use in the foreseeable future.

Echols doesn't see this consent decree as a success because a time line was only required for the northern portion of the county, which was designated as a "priority area." The consent decree told the county to clean up notoriously leaking sewers by 2020, but only two deadlines were established. The first was development of a hydraulic model that could be used to better understand how the system works and thus how to best target repairs. This model was not required to be completed until 2017. The second deadline was the end of January 2020, when presumably all necessary work was to be completed.

This sparse implementation schedule provided few opportunities for Echols weigh in: "I knew that I couldn't really do anything until 2020." She would have to wait until that date for formal comment. The work wasn't completed by 2020, and water quality standards had not been met. The South River Watershed Alliance submitted a formal complaint with the judge claiming that EPA and the county had not met the "diligent prosecution standard." In a legal ruling that strains logic, Judge Grimberg ruled that the consent decree was illegal because it did not contain deadlines required by the Clean Water Act, and EPA and the county had met the diligent prosecution standard. Essentially, the judge determined that EPA and the county tried. "The Clean Water Act doesn't say anything about trying," Echols tells me, an opinion not successful legally but that certainly wins in the court of common sense. The South River Watershed Alliance appealed the decision to the Eleventh Circuit Court and are awaiting a response.

The administrative response to a failed deadline is simple: push back the deadline. The consent decree was modified to 2027. In southern DeKalb County, despite Echols protests, no time line was issued in the modified decree. The consent decree listed a hundred or more known sites with leaks. Many of these were so egregious that they had been fixed before the issuance of the consent decree. "I have never seen a consent decree that applies to only one portion of a jurisdiction," Echols said. It's clear that the portion that it applies to happens to be overwhelmingly white.

While waiting for the legal challenges to play out, Echols has been engaging in other ways with various agency heads and judges. She walked out of a meeting with the EPA Region 4 administrator when he gave a presentation on how EPA is addressing environmental justice but didn't mention south DeKalb County. After EPA Region 4 refused to put a deadline in the modified consent decree, she raised the issue with the current EPA administrator, Michael Regan, and with Attorney General Merrick Garland. She wrote letters to Judge Duffy who was overseeing the consent decree. "He told me not to write them as they were ex-parte communication and that he'd have to publicly post them. I'm still going to write them," Echols says. Regarding the state regulatory agency Echols concludes, "if you don't raise hell with the Georgia EPA they don't respond."

When Echols talks about the issue of race, she negotiates differently depending on the audience. "The judges get pissed off when you bring up race," she says, but in theory the Clean Water Act should protect everyone equally if carried out fully. At the same time, clear discrepancies in water quality, land use decisions, infrastructure investments, and lived experiences of residents are obvious to anyone who cares to look. The increased attention to disadvantaged communities today must be both encouraging in theory and maddening in practice. As Echols explains, "You see it, and people don't want to talk about it . . . but there it is."

Different treatment from the regulators over leaking sanitary sewers is not the only discrepancy this area faces. The South River Federation got its start when a neighborhood association protested the expansion of a landfill that had already met its permitted capacity. The neighbors were successful in stopping that particular expansion, but that wasn't the end of landfills. South DeKalb County, and the South River watershed, is home to the two active municipal landfills that serve the county. "Environmental degradation is a game that plays out on the South River time and time again," she says. This level of agitation, advocacy, and coalition building would be impressive for any leader of a watershed alliance. It's notable that as a Black woman she has not pulled any punches. "A Black person heading an environmental group in Georgia is an enigma. We're the only one. Before us, there were none."

TAKE ME TO THE RIVER

While fighting these regulatory battles, Echols is also trying to sell the river to the public. "It's hard to generate interest in the face of negative headlines." One relatively recent win for the river was the construction of a trail and access point at Panola Shoals, a portion of the river that flows over granite shoals and creates shallow pools (figure 10.2). For someone wanting to cool off, the sand bars next to the river provide an accessible dose of natural beauty. One summer many people took advantage of this tranquil alternative to a city pool. Stories with headlines like "Frolicking in the Polluted South River" were published in local papers with pictures of people cooling off from the hot Georgia summer. "They almost blame the river," Echols says, "as if the river is responsible for its problems."

Getting people on the river is paramount for Echols. In Georgia and in most other states, water quality standards are dictated by the "designated use" of a river. There are six categories in Georgia: drinking water supplies, recreation, fishing, wild river, scenic river, and coastal fishing. This designation is intended to reflect the current use, and in an ideal scenario, it defines the allowable limits for criteria such as pH, dissolved oxygen, temperature, and bacteria. Setting a designated use doesn't necessarily mean that the water will meet it. In fact, approximately 50 percent of rivers and streams—703,417 miles—are categorized as "impaired," which means that the water doesn't meet the standard of the designated use.[6]

10.2 Jackie Echols at Panola Shoals on the South River, a watercraft put-in site and an occasional cooling-off spot for residents of Atlanta.

Standards are important and provide some guidance on what actions need to be taken, which was the case with the consent decrees in metropolitan Atlanta. Without them, there would be no impetus toward cleaning up the rivers, and judges would not be able to issue consent decrees mandating specific actions. For most sections of South River, the lowest level of use, "nonconsumptive fishing," has not been consistently met. This hasn't stopped Echols from pushing for a change to "recreational" use, which carries more stringent standards. This is the only way Echols can hope to hold the regulators accountable to address the ongoing pollution sources.

Echols has had some recent success in this regard. She submitted the formal paperwork requesting that all forty navigable miles of South River be designated recreational. Last year, the GAEQD accepted a portion of the request and changed the designation of the lowest thirteen miles. Echols swiftly resubmitted the paperwork for the other twenty-seven miles. She has not been provincial in sharing her strategy. In no small part due to her interactions with other watershed organizations, the state received an unprecedented number of requests to revise 1,600 miles of Georgia rivers in the last triennial review. The state is working its way through the backlog.

Getting people on the river requires more than just assembling a flotilla of kayaks, something that Echols and others do several times over the course of the year. Echols and her organization have been installing access points along the river. Stonecrest and some other towns have been supportive, but others, such as DeKalb County, read the writing on the wall and were initially skeptical. "The DeKalb County CEO at the time said, 'I'm not sure I can support this.' I told him I don't need his support." DeKalb County even posted signs at Panola Shoals threatening to arrest anyone using the river. Those signs have since come down, and the county seems to have accepted the idea. The county is procuring an in-stream trash trap that should capture some of the trash that remains a problem.

On the positive side, there is money for enhancing recreation. "There's no money for water quality, but there is money for recreation," Echols tells me. One jurisdiction told her that their matching funds had already been appropriated for the year; she tried again the next year and got the same story. Echols talked with them and said that she wouldn't accept the same explanation the following year. The municipality said they had no ownership of a proper take-out location. Echols paddled the river and found a perfect take-out spot. She then did a property search to identify the landowner and cold-called the owner. After several calls without a response, she went to the owner's house. The woman had many excuses for not selling the land, but after a few conversations she eventually agreed to sell 1.8 acres. Echols found the money, the municipality went along with the application for state funding, and the take-out was installed. "I try to remove the barriers," Echols explains.

HOPE RISING FROM NEGLECT

Fighting these battles on numerous fronts keeps Echols busy. Knowing she is an avid kayaker, I ask if she has any favorite places she likes to visit. She struggles to think of one, and then tells me she doesn't really have time to travel for that. She seems content to paddle the South River a dozen times a year with people who may be seeing it for the first time.

The list of injunctions, reviews, hearings, and public testimonies has my head spinning. Echols tells me that it's a challenge for her to keep it all straight as well. She was asked to write an article for a new environmental journal but had to push it back six months. "I don't want to say I can do it and then not come through." In a gracious manner, she tells me that the thorough accounting she is providing for me is helping her sort out the salient details. Many of the threats affecting South River seem to be coming to a head at this exact moment. I feel a quickening and a sense of purpose that kicks my nagging sense of powerlessness to the curb. It's eye-opening and fitting that this empowerment is emanating from this slender middle-aged Black woman who is driving me around in her Subaru.

Navigating the administrative and regulatory landscape that governs our streams and rivers is a challenge rivaling the Class 5 rapids on the Chattooga or the Chattahoochee. The framework of designated uses, water quality standards, impairment listings, and TMDLs to address pollutants has become a medusa-like monster with different programs addressing separate aspects of this process. Each assiduously carries out the requirements of their specific part, even if their part doesn't lead to success. At a recent meeting organized by the Georgia Environmental Protection Department, Echols presented her work in South River, and thirty regulators seemed desperate to find their way through this morass.

"What I try to get over is, 'Look, my goal is not a gotcha. My goal is how we can improve the river.' Not to just state what's in the TMDL—but be able to share how it's been effective," Echols says. "The effectiveness will be what builds support in communities. How is the TMDL improving the rivers? If not, why do it?" For anyone who has spent time in government bureaucracy dealing with the impaired rivers and streams, this sentiment is like sunlight breaking through a cloud bank. "Gotta believe that the river deserves to be better," Echols says. Coming from this most unlikely of action heroes, this belief is contagious. I feel its power, fueled by tenacity, focus, and a powerful sense of justice. It contains the force and energy needed to improve this beaten-up river and the lives of those living in its watershed.

CHAPTER 11

DAM REMOVAL AND COMPLICATED HISTORIES

Unfinished Business on the Elwha River in Washington State

WHERE THE RIVER MEETS THE SEA

Mike McHenry takes me down to the beach where the Elwha River dumps into the Salish Sea. We're at the northern edge of the Olympic Peninsula, the outline of Victoria Island in Canada is visible from this most northwesterly perch of the continental United States. This wide beach is strewn with logs and root wads that have been weathered by river currents, waves, and the wind. The land on which we stand, two hundred yards beyond the former coastline, is entirely a product of the removal of the dams. Forty-five million cubic feet of sediment that had been stored behind the dams made its way down to the ocean, and some portion of it settled to form this beach (figure 11.1). "We planted a few things but realized that the seed bank comes back pretty quickly," McHenry says, as we stare west toward the distant Pacific Ocean. It's a disorienting feeling, an impermanence brought on by the very newness of the land underneath our feet. The recently formed beach appears to have been here for hundreds of years, the low scrubby vegetation reminiscent of that found along similar coastal beaches on the peninsula.

Projecting the movement of this major source of sediment was the focus of intensive modeling. How much would be mobilized, how quickly would it move through the river, and what impacts would it have on the fish in the stream? These were all questions of major importance. Immediately after the dams were removed, the river would not be a habitable environment for fish. "We knew the levels of suspended sediment would be high enough to kill fish. We decided to use existing hatcheries to preserve genetic diversity," said McHenry. As the modelers

11.1 The newly formed beach where the freed Elwha River meets the Salish Sea.
Drawing by Claramae Hill.

had predicted, suspended sediment levels eventually dropped. "The sediment modelers nailed it," he said. Three years after removal, sediment returned to low background levels.

Although the river was the focus, this restoration of the river has benefited the nearshore coastal environment as well. This was something of a surprise. Shellfish populations have increased with this addition of sediment. This is important to the Elwha Klallam Tribe who have lived here at the mouth of the Elwha River as well as in neighboring coastal settlements along the northern side of the Olympic Peninsula for more than five hundred years. There are other beneficiaries as well. McHenry points out the holes dug by sand dwelling bees that are visible in the gaps between the beach pea vines that surround us.

We look back toward the old coastline at a narrow inlet that has formed behind the newly created beach. "This estuary was completely unplanned for," McHenry tells me, but its benefits have been significant. Before the disgorgement of sediment held behind the dams, salmon smolts swam from the fresh water of the river directly into the saline waters of the Salish Sea. Now they can spend time in this estuary to equilibrate. "Some salmon species need that more than others," McHenry tells me. The birds seem to love the area too. This area was not the focus of the restoration, and beach did not exist as it currently does, so prerestoration wildlife surveys were limited. Consequently, it's difficult to know the full extent of the benefit to wildlife. Having studied the river extensively for nearly thirty years, McHenry seems pleased by the multiplicity of ecological benefits; the predicted ones confirming our scientific understanding and the others emerging unexpectedly, providing glimpses of an ecological wholeness at the periphery of our perception.

The connections between the land and the ocean are palpable in this corner of the country. The water cycle, often seen in depictions of moisture moving along curved arrow paths, is made real here to all our senses. I imagine rainclouds in the mountain peaks only a few miles inland, smell the rotting carcasses of salmon that returned to spawn, and hear the river rushing through the canyons that were ideal dam locations. Here on the beach, with some imagination, I can taste a certain saltiness in the fog that covers the coastal landscape every morning until the sun burns it off around noon. These connections imbue the landscape with a sense of verdancy. Neither soft like pastoral New England nor harsh like the arid desert of the southwest, the Olympic Peninsula pulses with a sort of liquid energy. In this area, water envelops you in various states; it is as though this water cycle is juiced by a triple espresso shot. The energy is fueled by rain that falls in copious amounts, and it is most majestic and powerful in its clear and cold rivers.

In the Elwha River, these feelings are magnified tenfold. Removal of the dams combined with preservation of the headwaters created conditions for a level of restoration unseen in most rivers. It's an opportunity to undo the impacts of a century of exploitative management, to reestablish native fisheries, and to reverse the wrongs done to a native community. In essence, it is an opportunity to restore something fully. Most projects result in incremental improvements, but this project has the potential to put it all back together. Restoration of the Elwha River, a long time in the making and not yet complete, involves everything and everyone.

ADDRESSING A CRIMINAL ACT:
TWENTY-FIVE YEARS IN THE MAKING

We're back in the truck and heading upstream to meet McHenry's staff who are already at the sampling site on the river. This visit has loomed large in my mind for the past year. We've spoken on the phone a couple of times, but because of COVID it's taken this long to get out here. This is just the recent history; this river has been scouring eddies and leaving deposits in my subconscious for nearly three decades. Now here, my recollection of the area, out of date and tethered to memories thirty years old, collides with the immediacy of the present. I ask questions, seeking my bearings. McHenry is friendly and patient with my rambling inquiries. The restoration of the Elwha is a project that took twenty-five years to realize. Its story is contoured with complicated politics and evolving public opinion. I try to make sense of the intersections of biology, culture, and economics that have played out here. The Elwha is a story that is highly specific to itself, but it also contains universal lessons for those working to restore streams and rivers.

McHenry's responses are short and to the point. Like any good boss, he aims to be understood. When we left the office earlier, he held a quick meeting in the foyer where his staff checked in on their day's assignments, and he answered two calls from his staff on the drive to clarify their tasks. I ask him about the politics that eventually changed and allowed this project to happen. In an earlier call, McHenry had mentioned Senator Slade Gordon's opposition to the project and his potential concern that the removal of the Elwha might open the door for removal of the dams on the Snake River in eastern Washington. As the chairperson for the Interior Appropriations Subcommittee from 1995 to 2001, Senator Gordon held the purse strings when interest in removing the dams was building. I ask McHenry to elaborate on our past conversation, and he clarifies, "That was just my speculation. I don't know former Senator Gordon and obviously don't operate in those circles."

McHenry seems comfortable not operating in those circles. He has worked as the fisheries habitat manager for the Elwha Klallam Tribe since 1991. Prior to coming to the Olympic Peninsula, he worked in Idaho for a brief time for the state forestry department. He saw how political influence dictated management decisions and left that situation as soon as possible. Working for the tribe, he actually gets to protect the resource. His status as a nonmember doesn't seem to affect his ability to say what he believes needs to be said. McHenry has been a key player in the long and sometimes tortuous path that finally led to the removal of the lower dam in 2011 and the upper dam in 2014. The ecological, cultural, and emotional significance of this project cannot be overstated. "For the tribes, the dams were always a criminal act," McHenry tells me. Even though hatcheries were built to grow and release salmon, all eight natural salmon runs were decimated. The salmon populations were put on life support that would continue for eight decades.

An unspoiled area was saddled with what could only be described as an ecological travesty. The area has received increasing levels of protection, first as the Olympic Forest Reserve in 1897 and later as the Olympic National Park in 1937, but no one could consider this area unspoiled with these dams in place. A river with an otherwise pristine and protected watershed was tamed by these dams. One of the two dams was located within the park, the other just outside of it. They had been built to power lumber mills that would deforest much of the northern half of the Olympic Peninsula. If that bargain wasn't offensive enough to the river, a water diversion pipeline was built to supply seven mills in Port Angeles, which also drew significant volumes of water. Today only one mill is still in operation, processing logs from the forest outside of the Olympic National Park into packing grade containerboard.

The shape of the decline of the eight salmon runs was probably different for each species, but the trajectory pointed in only one direction. McHenry tells me that the decline was never documented. It's not surprising that no one documented such a precipitous and shameful demise. Some runs, such as the spring Chinook and the summer steelhead, were eliminated immediately. Other species faced a slow-burn decline. In the end, the local sockeye run was destroyed, the pinks and chums were reduced to barely measurable levels, and the bull trout were soon placed on the endangered species list. The Chinook, Coho, and steelhead were kept alive through hatchery operations. When these runs returned, the only place they could go was to the lowest four miles of river, which had none of the appropriate riverbed material for spawning. These three species were put on the equivalent of an end-of-life breathing machine, hanging on thanks to the costly ongoing efforts of the hatcheries. Between 1950 and 1980, even the lowest section of the river was subjected to dredging. This finished off any remaining habitat any remaining fish might have used.[1]

Arising from such calamitous actions came the chorus for something better. Such is the draw of restoration: making amends for the stupid actions of others and creating a new story. Removing the dams not only restored fish runs but also righted a wrong to a people whose culture had been negated and ignored. It was restoration in the most expansive sense of the word—ecological, cultural, historical, and for many spiritual—all wrapped up in one massive undertaking.

The project gained traction in the 1980s when the Federal Energy Regulatory Commission notified the dam owner, as well as the tribes, the park, and the state that the dams had no active license. Fish passage would be required for the dams to continue to operate legally. This precipitated a fuller assessment of the threatened and endangered status of the salmon runs through an Environmental Impact Statement (EIS). Fish passage would not be sufficient for recovery of the salmon populations. This assessment through the federal EIS provided the regulatory muscle to begin the planning process and the larger political process for securing funds to implement a project whose total costs would swell to $324 million, a level previously unseen for a river-restoration project.[2]

Shortly after McHenry started working with the tribe, the Elwha River Ecosystem and Fisheries Restoration Act was passed in January of 1992. This act authorized and directed the Secretary of the Interior to begin planning for "removal of the dams and full restoration of the Elwha River ecosystem and native anadromous fisheries." A brief but important line in the act (102–45), a number McHenry knows by heart, states "subject to the appropriation of funds therefore." This last line required majority support in Congress, and critically, the political support of local congress members and senators. Maria Cantwell's defeat of Senator Gordon in 2000 by a mere 2,229 votes out of 2.5 million tilted the balance of power in the state and in Congress. It meant that this project would receive full support from key local representatives.

The process involved years of public outreach and represented a unique balance of power in which the feds, the state, and the tribe were more or less equal powers. By the early 2000s, Congress was appropriating smaller amounts toward the project, but not the amount required to begin the project. "That was the hardest part of it," McHenry relays. "The drip, drip, drip nature of the support. I wasn't sure if it would ever happen."

President Obama's stimulus bill and the need for "shovel-ready" projects provided a chance to finally infuse the project with the massive resources it needed. The project was eventually competed, and as McHenry says dryly, "the Elwha had its fifteen minutes of fame." For McHenry and other fisheries biologists, the story continues beyond this moment of exhilaration and relief that came with removal of the dams. Management and monitoring fish populations is, by its nature, ongoing and not conducive to a satisfying, Hollywood film variety resolution. McHenry has responded to hundreds of journalists seeking that feel good story, something to combat the ecological trauma that many find themselves reporting. McHenry tries to oblige but doesn't seem ready to declare mission accomplished and retire to watch depictions of ecological restoration in the comfort of his home. His wife frequently asks him when he's going to retire, but he has no plans to do so. The story is still being written, and I suspect he wants to witness the conclusion in the fish and smolt data his team collects.

McHenry believes removal of the Elwha dams presents our best chance to return a river to its wild state. This river has the potential to become the first river to see the return of natural salmon runs after decades of hatchery dominated fisheries. Wild runs of Coho and summer steelhead have already made dramatic recoveries. Bull trout, present in small numbers upriver before dam removal, have reestablished anadromy since the removal and are increasing in size, reaching ten pounds. Small numbers of Chinook, the prized king salmon, have begun to use the upper reaches now available to them. McHenry and his colleagues have employed techniques ranging from eDNA to snorkel surveys to gauge their return.

But McHenry reminds me that the story is not finished. Pink and chum salmon are at very low populations and are at risk of collapse. The Elwha sockeye run was extirpated by the dam, although a landlocked form, the Cockinee, resides in Lake Sullivan, a lake that feeds one of the main tributaries of the Elwha. The sockeye are in a "natural recovery boat," he tells me. Occasionally salmon stray from their river of origin and enter other rivers. There is potential for that to occur with the sockeye. There have been discussions of bringing in neighboring stocks of pinks from other rivers should funding materialize.

The hatcheries still produce steelhead, Coho, and Chinook, and many would like to see them continue to do so. When the dams were in, the hatcheries turned out thousands of salmon that eventually returned to the short four miles of stream available to them. In that situation, it made sense to harvest as many as possible because significant levels of spawning habitat weren't available. "As soon as we mess up a watershed, we move in the opposite direction and put in hatcheries," McHenry tells me. Take all you can because the hatchery will make more. He says that the harvest methods used for a hatchery supplied fishery won't work for wild runs. This statement is unpopular with the tribe because the methods he refers to are considered to be traditional by tribal members.

How various groups will enjoy any future bounty has not been resolved. The issue is timely and also controversial. After eleven years of a fishing moratorium, calls for opening up the salmon harvest are getting louder. A tribe whose culture and way of life was upended by the dams, a power grab in the most literal sense, is now waiting for some relief nearly a century later. The many scientists working on the river are trying to figure out when that can happen, but there has never been a project quite like the Elwha.

MEASURING RIVER HEALTH ONE SALMON SMOLT AT A TIME

The floating platform that holds the screw trap is secured to the bank closest to us. The current rages this time of year, supplied by snowmelt from the mountains as well as recent rains. The river at this spot is more than 130 feet wide and gathers force as several braids connect only a few hundred yards upstream. McHenry implores me to hold on to something solid the entire time I'm standing on the pontoon-like structure. "You don't want to go in," he tells me with his characteristic brevity. A screw trap is comprised of a barrel five feet in diameter that spins with the current. It ends in a funnel that directs smolts into a small, rectangular, grated pen where the salmon rest as river water flows through (figure 11.2). The entire device is supported by metal walkways and floats underneath and is cabled to both banks by several thick steel cables. I step on and hold onto one of the cables and try to stay out of the way.

Mel Elofson and mcKenzi Taylor, members of McHenry's staff, get to work identifying the smolts that have been detained on their trip down to the Salish Sea. The purpose is to understand how many of each species are out-migrating. Comparing these numbers to the number returning in future years will help them understand the strength of the runs and any other pressures that might be affecting the fish in the ocean. Taylor and Elofson scoop the smolts from the holding area with small nets and place them in camo-patterned, five-gallon buckets. Taylor begins to pick the fish out of the bucket individually and examines them.

11.2 Mel Elofson shows off a salmon smolt caught in one of the screw traps he customized and installed.

She rapidly turns the fish over in her hands, examining the par marks along the side, noticing the tail fin, and inspecting the shape of the anal fin. "Coho" she tells Elofson and then throws it into another bucket. "Chinook" she says with the next and tosses it back into the river. Taylor calls out "Mykiss" many times, which is not surprising because a hatchery less than a mile upstream releases this species every year. *Oncorhynchus mykiss* expresses two different forms. When confined to freshwater, as was the case before removal of the dam, they are rainbow trout. The ones that Taylor is seeing are the steelhead form, or rather they will be once they return three years later and are up to forty pounds heavier.

Taylor quickly points out some of the differences to me, but it's an avalanche of unfamiliar fish body parts. She cross-checks four or five indicators in less than five seconds as she turns the small fish in her hand. When I begin to think I might be able to make an identification, I learn about the occasional hybridization between rainbow trout and cutthroat. The characteristic red slash on the cheek of the cutthroat can sometimes be seen on the rainbow.

I quickly dispel any notions of getting this and am in awe of their field knowledge. Occasionally things get confusing for the fish biologists as well. Hatchery fish can be confusing in different ways. Frequently exhibiting odd injuries, scrapes and lost fins, the hatchery fish usually stand out. "You don't see those sorts of injuries in wild fish. Hatchery fish see all sorts of abuse in that environment," Taylor says.

They are separating the Coho to get a sampling efficiency rate for their screw trap. The smolts are dyed with an orange-brown dye and then are released about 500 meters upstream. The percentage that return into the screw trap provide a good indication of the capture efficiency of that trap. Today Elofson is returning fifty-three to the river. Coho have been doing well. Indian Creek, a low gradient tributary to the Elwha, has been open to the fish for ten years and provides ideal habitat for this species.

Before the dam removal, 98 percent of the Coho were hatchery origin fish. The smolt that the hatchery released would migrate to the sea, but upon returning they would not be able to spawn upstream and complete their life cycle. In a futile attempt at reproduction, these hatchery fish oftentimes returned to the hatcheries at river miles two and four. They would key in on the chemistry of the hatchery water, Elofson informs me. The fish had lost their anadromous behavior; they were salmon in shape and form but not in function or spirit. These fish needed help. In 2012, after removal of the lower Elwha dam, McHenry and his team began collecting surplus Coho adults and transplanting them upriver on Indian Creek and Little Creek, tributaries now reconnected to the mainstream. They moved more than 3,200 adult Coho this way over a ten-year period, and the effort seems to be paying off.

Summer steelhead, the anadromous version of Mykiss and the star performer to date, come up regularly in Elofson and Taylor's tally. This may be because a healthy population of rainbow trout existed upstream before there was a dam. It's thought that these rainbows immediately returned to anadromous behavior. Due to the success of this species, the tribe has recently lowered their target of hatchery released smolts from 175,000 to 30,000.[3]

I ask Elofson about his history with the river and the dams. His grandmother lived not far upstream on a homestead, and in his youth the tailways of the dam were a natural place to fish and generally to hang out. He didn't know the river before the dam was built, but his grandmother passed her memories on to him. "I remember my grandma [saying] that the pinks were so thick in the summertime that you could walk across the river," he tells me. She also told him about seeing the dams being built. Living so close, his grandmother witnessed the construction of the dam and the thousands of fish dying in the pools as they tried to get upstream.

Elofson points out how the river has changed form since the dams have come out. Before dam removal, the river bottom was starved for smaller cobble and fine materials. Bowling ball–sized rocks made up the stream bottom, material much too large to be used for spawning. Since removal, the river has chewed away certain banks, and in the process multiple channels have been created in several locations. This seemingly destructive process mobilizes the smaller material that generates a riverbed suitable for spawning for both the salmon and Pacific lamprey, which also use the Elwha. In a supporting role exceeding their size, juvenile lamprey constitute an important food source for salmon in the river. In their larval form, the fatty tissue of lamprey is forty-five times richer than salmon tissue. Increasing numbers of lamprey, spawned in the sandy bars of the tributaries, allow young salmon to put on weight that improves their survival. When the salmon return, their spawning success is greater and the numerous species that depend on their biomass thrive in turn. Another species, similarly linked to the sea, responds to restoration of the river and offers its progeny to support the return of the salmon.

The trees that are taken down by the erosive force of the river create logjams that capture the smaller cobble and finer sediment. Before removal and immediately after, McHenry's team installed more than thirty-three log jams to jumpstart this process. These engineered log jams were a significant undertaking that required large machinery to key the logs into the banks. Now freed, the river is doing this on its own. A large log jam on the opposite bank testifies to this process, which is integral to the health of mountainous western streams.

The river has exerted its force on Elofson physically. Now sixty-three, he has acted as McHenry's right-hand man for much of the ongoing monitoring work. He has tweaked, modified, and installed several of the screw traps that are deployed in key locations in the watershed. A few years ago they conducted a gravel survey of the riverbed as part of a broader habitat survey. While floating on the river, the raft his team was in, loaded with 2,000 pounds of gravel, veered toward a log that hung over the river. The impact cracked two ribs and cracked his sternum. Had the log not eventually broken "it could have been worse," Elofson tells me unceremoniously. He stretches uncomfortably as he describes this harrowing tale. Cracked sternums heal slowly for those in their sixth decade of life. It's clear that he's still feeling the effects.

Standing here on the bank and seeing the evidence of salmon returning to the sea, it's easy to believe that the hard work has been done, these smolt will return in large numbers, and the bounty of this river will return. Given the eleven-year moratorium on fishing, it's high time to enjoy this bounty. Elofson hears this a lot from his fellow tribal members. Some members say that the elders have suffered immensely from the fishing moratorium. The loss of this cultural touchstone, magnified by the isolation that COVID brought, has been too much. Although sympathetic to these challenges, Elofson and other fisheries biologists are reluctant to restart the harvest, at least in the form that existed before the dam removal.

Things have changed for the tribe as well; there were fewer options twenty years ago. The salmon helped pay a lot of utility bills, but many of the anglers are now too old for that. The younger ones have moved on to the shellfish harvest, which is much more profitable. Other jobs are available now too. The tribe is building a new casino and a hotel.

If patience is what's needed, Elofson seems to embody that quality. I asked him if he has taken part in the intertribal gathering that includes a two-week ceremonial canoe trip. In late summer, the Salish tribes around Puget Sound and Vancouver Island travel to all of the neighboring tribal villages and pick up fellow paddlers along the way. People switch up seating and reconnect with acquaintances. The year before COVID-19, the participants went all the way down southern Puget Sound to Squaxin Island. A few years earlier, the flotilla crossed the Salish Sea and paddled three-quarters of the way up the eastern side of Vancouver Island. "I might when I retire. I've got two more years," Elofson says. "Mike kind of relies on me here, so I can't take that much time off."

Imagining a future bounty that could sustain both a people and an ecosystem is intoxicating. The payoff from restoration is nearly palpable but still eludes our collective grasp. Rather than a particular solution or action, it requires patience and restraint. I know Elofson wants this payoff that drives all involved in restoring our rivers, but he leans on what he knows through his work. He has a detailed knowledge of the evolving habitat of this river and the salmon that are using it. His data indicate that this bounty is on the horizon, but the river may not be ready to provide that bounty yet.

HOW TO CATCH A SALMON

McHenry has taken me out to a spot on the Elwha where Roger Peters, a colleague from the U.S. Fish and Wildlife Service (USFWS), is undertaking a study in coordination with the Lower Elwha Klallam Tribe. This is not an uncommon occurrence; after the dam removal, the Elwha has become the most studied river in the country. The Dam Removal Information Portal, a database that tracks research, indicates that the Elwha has been the subject of fifty-seven studies since removal of the dams. These studies cover all aspects of physical, biological, and water quality. Considering only the subset on fish research, a significant data set extends back several decades: eDNA water samples have been used to determine how far upstream each species has spawned, and direct snorkel floats have measured numbers of species in different sections of the river. The river has been well studied, but Peters' study is unique in that it is attempting to understand the mortality rates of various fishing methods. Peters and his team are catching fish to determine how likely it is that one of these caught fish could be released and still survive.

McHenry and the Elwha Klallam Tribe have sponsored this study to better understand how fishing could be reopened on the Elwha. Eleven years into a fishing moratorium, both tribal members and recreational anglers are eager to realize some of the benefits of this monumental dam removal project. The decision to allow some fishing has not been made, but there is great interest in allowing the harvest of some fish. The decision of how to fish and which fish to harvest involves complicated considerations that balance cultural and economic needs with evolving salmon population dynamics that, despite the extent of research, are not completely understood.

The sun has cleared the morning fog, revealing wispy cirrus clouds high in an otherwise blue sky. Peters's team is floating a large drift net down this section of the river. Two younger men equipped in neoprene waders hold the ends of the net on either side of the river, and Peters holds the center as he floats downstream (figure 11.3). Several other people are milling about on this wide gravel bank, and a videographer is documenting the work. Apparently there are wider audiences for this study. The net is weighted to hit the river bottom, and it will capture salmon as they try to escape. The informal chatting among the large crowd on this radiant day feels festive. As Peters's team makes repeated floats down this

11.3 Roger Peters (center) and his staff float a drift net down a section of the lower Elwha as part of a collaborative study between the U.S. Fish and Wildlife Service and the Lower Elwha Klallam Tribe.

same stream section, I learn that this study will influence upcoming decisions about the harvest. Both tribal and sports anglers will be affected by the results.

Peters has developed a methodology that attempts to mimic traditional fishing methods. When Vanessa Castle, one of McHenry's staff, approaches me to be sure I understand that tribal members have been suffering during the eleven years of the fishing moratorium, I realize that this study is about more than just the salmon. There is unspoken tension because these traditional methods are now subject to a scientific study and, by and large, this assessment is being carried out by a federal agency. As Olympic Park superintendent Rodger Conner stated in 1982: "To characterize the management of wild fish stocks as controversial would be a considerable understatement, especially on Washington's Olympic Peninsula."

Peters is a long-time fisheries biologist for the feds, a tribal member, and a commercial fisherman. His tribe, the Squaxin Island Tribe, has landholdings and historic fishing and shellfish rights in the seven southernmost inlets of Puget Sound. Peters commercially fishes these inlets in addition to his full-time job. Understanding the desire from both tribal and recreational anglers to allow fishing once again, he says that the USFWS is trying to get in front of this issue. They met with a group of Elwha Klallam fishermen to get their support for any future harvest regulations. He says that the turnout was disappointingly low. Those who came expressed interest in trying gill nets again, the traditional method most familiar and effective for tribal fishermen.

Peters says he understands because this was the method used before the dams were removed. Although the dams prevented any natural runs of salmon, two hatcheries still turned out Coho, steelhead, and Chinook, and the hatchery salmon returned to their birthplace of concrete sluiceways. Peters describes the harvest of Coho at this time as fairly robust. Tribal fishers used drift nets, set nets, and even handheld nets during these runs. Much of the fishing was for subsistence, but many fish were sold for money. Most of the tribal members lacked the resources for expensive fishing boats, and these techniques were inexpensive and effective. The sustainability of this fishing was questionable. As long as the hatcheries continued to pump out smolts, the returning fish could be harvested without concern for long-term population health. The fishery was fully detached from the natural context of spawning.

This situation was in no way unique. Most of the fisheries in the United States are supported by hatcheries in this manner. Hatcheries deliver fish to waiting anglers in much the same way that pheasant breeding farms release the birds to expectant hunters that await them with gun barrels pointed skyward. In this way, the differences between our hatchery fueled fisheries and our vertically integrated poultry industry are mostly related to where the production is realized. For chickens, we create high-tech warehouses to facilitate this output. For fish, we use our rivers and oceans, as if they served no other purpose than to maximize production for human consumption.

Currently, some of the native salmon runs of the Elwha are improving, but the picture is complicated. The summer steelhead are extremely strong, but the Chinook have not yet shown strong returns to the native runs. "Eleven years after dam removal, we're only in our second generation of Chinook," Peters tells me. The 2019 cohort of 571,00 Chinook smolts have yet to return in significant numbers.[4] Chinook typically spend four or five years out at sea, and Peters is hoping for a strong run next year. But this cycle means that a return to historic

Chinook runs may take several cycles over decades. If the most recent cohort somehow performs poorly, it could be several additional years before a recovery. And should the catch of any hatchery fish include a significant incidental catch of wild run Chinook, the pace of recovery could be slowed or stopped.

Gill nets have been a flash point in the numerous conflicts related to fishing regulations. Despite their cultural relevance for tribes, it is indisputable that they cause high mortality to the fish caught in them. This is relevant because, as Peters tells me, "there is the potential for a harvest of hatchery fish, but we want the native origin run to continue to build." The nets don't distinguish between the hatchery run and the native run, and most important, the mortality of net fishing in general is about 50 percent. Any Chinook Peters's team catch in their study will be taken to the hatchery to see if the fish will still spawn. His study is using traditional gill net techniques, and modifying certain ones, in an effort to quantify the mortality rate. If the mortality rate can be lowered, gill net approaches could be allowed.

Some modifications have shown promise elsewhere. The Nisqually Tribe, also located in the southern Puget Sound, is using nets with smaller openings. These are less likely to catch the salmon's gills and cause mortality than nets with larger openings. For his research, Peters is trying shorter drift times, less than five minutes, that would allow a drift netter to quickly release a native origin salmon. Set nets up to 150 feet long are secured on one end in the center of the stream and traditionally are left overnight and harvested in the morning. One component of the study involves installing a set net for a shorter period with continuous monitoring.

The approach that involves the lowest incidence of mortality is "hook and line," what most people think of as recreational fishing, which averages a 10 percent mortality rate. "Hook and line" seems to be the most palatable option to those responsible for restoring the wild salmon runs of the Elwha, but Peters says they need to look into this further. Commercial recreational guides who would be quick to take advantage of a limited hatchery-based harvest are adept in their craft. The potential to catch salmon in a restored Elwha would bring high-dollar customers from across the world. Some native origin salmon might be caught multiple times. Peters says they don't yet understand the mortality in that situation but are working on some potential studies to determine it.

I ask Peters if he feels a unique pressure as both a scientist and a tribal member. "Not really," he says, "I work for the fish." As an employee of the Department of the Interior, he takes the goal spelled out in Congressional Act 102–45 in 1992—"the full restoration of the Elwha River ecosystem and native anadromous fisheries"—as a clear directive. "We have a huge responsibility for the Elwha dam removal project to be successful." Success, he says, is when we have fishing in a sustainable manner. "We haven't succeeded yet," he says. "If we don't, it may be impossible to do this again in other places." Peters explains that there are many other places where dams are outdated, unsafe, or have otherwise lost their license. The eyes of the restoration community and those that fund it are clearly on the Elwha.

The issue could not be more timely. Peters explains that the state, the tribe, and the feds (Olympic National Park) plan to meet in January to discuss the potential for opening the fisheries for harvest. It's unlikely that gill net methods will be allowed, and he describes the reason for this somewhat indirectly: "The hindrance to tribe and recovery is that traditional gill net fishing will provide very few opportunities to harvest." This parallels McHenry's statement about allowing traditional methods. After eight decades of ecological and cultural injustice, it's difficult to come out directly and say that the traditional fishing methods for the Elwha shouldn't be allowed. But as trained scientists, Peters and McHenry understand that this is what the science is telling them.

I ask Peters if the presence of the hatcheries is helping or hindering this recovery. The release of thousands of smolts that come back as fat salmon provide a much-needed economic resource. Salmon are a cultural touchstone for both the tribe as well as the Pacific Northwest as a whole. This artificially juiced bounty calls out to be harvested. Just as the typical produce shopper can't walk past grapefruit-sized apples rendered aesthetically perfect thanks to a bevy of pesticides and subsidized irrigation, most people aren't concerned about whether a salmon fillet started as smolt in an unimpeded and healthy river or came to fruition in the sluiceways of a fish hatchery.

These hatcheries have served a purpose, Peters reminds me. The Elwha stock of Chinook probably would have become extinct if fisheries managers hadn't captured them prior to dam removal and reared them in hatcheries. The recovery of others, such as the summer steelhead, may have been hastened by timely hatchery releases. I ask Peters what would happen if the hatcheries were closed today. "Some species like the steelhead and Coho would be fine," he says. Others, like the Chinook, "wouldn't be around without them." But the hatcheries aren't free. In exchange for cheap hydropower and developing watersheds, we pay for the operation of hatcheries. "Nothing in life is free," Peters tells me. "We are paying for every decision we make. The currency we're paying for the development and dams is the continued use of hatcheries."

Elaborating on the scenario of closing the hatcheries, he adds that we'd "lose harvest in the near term, but it's hard to tell for how long." I'm struck by the awkward position fisheries managers are put in by our society that craves abundance at any cost. The desire for maximum sustainable yield, the economic sweet spot that provides the greatest output, leaves little room for returning a wild native run. Yet we want both: the expensive and hard to grow heirloom tomato and the cheap and abundant Romas to grind into paste. In the case of the Elwha, we might have to make a choice.

FISH TALES AND FISH POLITICS

McHenry places me in the care of Philip Charles and Vanessa Castle, two Elwha tribal members who also work for the tribe. We're about half a mile downstream from where Peters was running his drift nets. Charles, a conservation warden in his fifties, has close cropped hair and is wearing a black ACDC shirt and sweatpants. His everyday outfit stands in contrast to several elaborate tattoos on his arms that highlight tribal imagery. Castle is a wildlife biologist in her thirties who works for McHenry in the fisheries group. She wears waders and has her long black hair pulled through the back of a white baseball cap.

None of us have an official task at the moment, so we watch Jacob Bengelink, one of Peters's staff, attempt to install a set line in a section of the river where the current is particularly strong. He is outfitted in a wetsuit and has rigged carabiners to a thick cable to ensure that he won't be swept downstream as he crosses the river. His hundred-foot-long set line has drifted in the swift current to the far side of the river and is entangled with some tree branches that extend into the water. Charles, Castle, and I watch Bengelink fight the current as he crosses back toward our side. Not being properly outfitted and therefore unable to help, we offer conversation as he struggles with the net. Castle appears to be annoyed about the study and says that several tribal members could have assisted in setting this up. Although they haven't used them in the past decade, most tribal members are familiar with this setup. She claims that the tribal members would have been more engaged if they could have conducted this experiment.

Castle brings up the issue of blood quantum when I ask about the requirements of tribal membership. She explains that the system was enacted by whites to make natives prove they were deserving of any help. "I hold strong and don't tell people," Castle says proudly. "I tell them I'm 100 percent in my heart." The requirements, however, pose some challenges to finding a spouse, she mentions. Marrying the wrong person could make your children ineligible for tribal membership. "It's not easy to date on the reservation," she laughs.

We begin talking about boats, and I mention a carved canoe I noticed behind one of the tribal buildings. Charles proudly tells me that he went on the last multitribe voyage down to Squaxin Island, a distance of more than 125 nautical miles. When they get there, each tribe conducts its own ceremonial songs and dances. The process can go on for days, and no one is in any rush to leave. I feel a twinge of jealousy for such a vibrant sense of belonging. Castle brings our attention back to the river, pointing out a salmon, a fleeting shadow from our perch. In their own ways, Charles and Castle appear fully committed to this connection, their sense of pride visible under the surface.

Bengelink has returned to our side of the river, frustrated by the tangled set net that won't extend properly. Although he's done this work all over the country, he's decided that this isn't a one-person job. He'll have to discuss this with Peters. While Bengelink gets out of his wetsuit, the four of us chat about fishing. Hearing that I'm from Wisconsin, Bengelink begins talking about his experience fishing for sturgeon in Lake Winnebago, notable for its sustained production of a fish highly susceptible to overharvesting. Charles mentions the fishing prowess of his "crazy uncle Floyd," who used to fish this section of river alone using an approach not too different from what Bengelink was attempting, and adds that he'd like to try to weave a net in the traditional manner with cedar bark material. Bengelink describes a recent trip to Iceland where the fishing rights on every river belong to the property owner. After hearing the price for taking a fish from the Icelandic owner, he decided on a less costly hook and release option.

In spite of our different backgrounds and the underlying tension related to the study, fishing is our shared language. Anglers from all backgrounds share a body of knowledge and a breadth of experience that breaks through differences that might otherwise divide them. People want to share their stories of bounty, and for the most part, people want to hear those stories. Harvesting skills that reflect an understanding of the fish and the river they inhabit are respected.

11.4 Vanessa Castle, member of the Elwha Klallam Tribe, holds a boulder that was probably used as a weight for a traditional fishing net.

Strolling along the river's edge as we talk, Castle lifts a large smooth stone shaped like an hourglass. She points out that this was probably used as a weight for a fishing net (figure 11.4). The immediacy of the rock and its connection to those who used it provide a data point more compelling than any that might be gathered with a range of high-tech equipment. This boulder, smoothed by rope and current, testifies to the connection between a river and a people. She holds the boulder a bit longer, seemingly comfortable with its weight, before she finally puts it back down among the others.

Fishing is also a point of cultural tension. Centuries of gradually improved management can cloud the fact that traditional harvest methods wouldn't present any threat to Elwha salmon stocks had the colonizers not dammed the river in the first place. The moratorium, as important as it is to the return of salmon, is the product of a tsunami of cultural and ecological impacts unleashed by the dominant white culture. The descendants of these European colonizers, now equipped with fly rods, also have a desire to take part in this harvest. For better or worse, we now share a resource that puts limits on us all. Evolution of the relationship from oppression to nation-to-nation negotiation is necessary and, when applied to limited natural resources, challenging.

Charles will be tasked to ensure that regulations are followed, and I ask if it would be possible to enforce any restriction on native run salmon. "Pretty hard to enforce," he tells me. He has no jurisdiction over nontribal members, who in many cases are hostile to his presence. The discussion shifts to a recent harvest on the Quilcene River, about forty miles southeast, that is also part of the traditional fishing grounds for the Lower Elwha Klallam Tribe. Castle shows a video of a river bubbling with an unfathomable number of salmon and tribal members scooping up the fish with common fishing nets. Even on a small screen, the video has the impact of an IMAX production. Each tribal member was allowed six salmon, and Castle says she brought along all the kids she could round up. Charles gently reminds her that children with the adults are not supposed to take six for themselves. The fishery on the Quilcene River is supported entirely by a hatchery run by the U.S. Fish and Wildlife Service. Perhaps even more complicated than the ecological issues, the connections between cultures reflect patterns of exploitation and dependence. To my outsider's eye, this dynamic seems to leave all groups uncertain and, perhaps, looking for a new arrangement in which coexistence is built on mutual respect and collective responsibility. This is an arrangement that requires an ecological abundance not dependent on hatcheries, which is as foreign to the current parties as it is to an outsider such as me.

As we leave the site, I feel uncertain about the return of wild salmon stocks in the Elwha River. The outcome that originally seemed entirely dependent on removal of these massive dams now seems more dependent on the decisions of agencies and the tribes, who represent only parts of the whole rather than the entire ecological and cultural cosmos. The ultimate feel-good restoration project that could unite people may just as easily divide them. For her part, Castle seems uncomfortable with what the study will recommend. "They'll let us hook and line, for science." she jokes with Charles. The dark humor hits with the weight of the boulder she was recently holding. If there is no mutually accepted process for managing this river, how can it possibly return to its former abundance?

A RIVER REWILDING

And when would the road be built? Nobody knew for sure; perhaps in a couple of years, depending upon when the Park Service would be able to get the money. The new road—to be paved, of course—would cost somewhere between half a million and one million dollars, depending on the bids, or more than fifty thousand dollars per linear mile. At least enough to pay the salaries of ten park rangers for ten years.
—Edward Abbey, Desert Solitaire[5]

When I asked McHenry if there had been any conflicts with the park, the only thing he mentioned was the road. The Elwha River Road allowed access into the park and to the former Glines Canyon dam site. In addition to a ranger station and several maintenance buildings lower in the valley, the road allowed access to the Olympic Hot Springs and two trail networks that extended into the park. Shortly after removal of the dam, however, the river washed out this only access into the park within sixty miles.

Repairing the road would be more complicated than just replacing a bridge or putting in a new roadbed. The river chewed away at the foundation of the road as it pushed toward the valley walls, turning parts of the former road into a new channel. Replacing the road would require placing substantial stone and concrete reinforcement in the active channel or tunneling into the valley walls. The tribe had protested the National Park Service's (NPS) initial proposal to rebuild within the channel and suggested that NPS construct tunnels into the valley wall to avoid building in the stream. NPS had balked at the potential costs that tunneling would involve. I asked McHenry what value the road provided, and he described it as allowing access for a handful of hikers. He also added that with the appointment of the new Secretary of the Interior, Deb Haaland, NPS recently began the process anew. A full environmental assessment would be conducted that would consider the comments of the tribe.

Edward Abbey details the exponential increase in visitors to Arches National Park following paving of the entry roads. Here the removal of the dams had provided a rare reversal in this steady march toward greater accessibility. I wanted to head up that road for a few reasons. I wanted to witness the destructive power of this newly freed river. I also wanted to return to the Olympic Hot

Springs, a favorite destination of mine during college. And I wanted to see the site of the former dam and reservoir with my own eyes. But I also wanted something more difficult to describe. Maybe I wanted, as Abbey had railed, to "be pulled out of [my] back-breaking, upholstered mechanized wheelchair, and onto my feet, onto the strange warmth and solidity of Mother Earth again."[6] I decided to spend two days and head up the valley, a ten-mile hike from the Madison Falls parking lot.

My memories, over thirty years old, were as hazy as the coastal fog that hangs over this stream valley. I was certain that being here would sharpen them, and the specific details of the landscape—the trees, the clear river, and the roads that wind into the park—would come into focus just as the lifting fog reveals these same features every afternoon. I could recall the route; traveling upriver in my mind. After passing the sign for the park, I drove up through the valley, gaining elevation along the way. The valleys are steep so vision is limited to the immediate watershed, and visitors can find themselves suddenly lost in this park if they stray off the trail. Although only five miles from a town, any indications of civilization are absent. Except for a network of trails and some established campgrounds, the land is wild.

In my memory there is one point where the road turns toward the river to reveal the 210-foot-high wall of concrete. Armed with my newly awakened environmental consciousness, this dam symbolized all of the ecological and cultural baggage that underlies the American economy. It was inconceivable how this, and its companion just three miles downstream, could have been allowed. It was also inconceivable to me at the time that this concrete mass could ever be removed. It seemed to be proof of the original sin of the European settlement of the country, implicating our way of life, from the paper pushed in tidy offices to the cheap energy that powered the endless electronic devices that simplified our lives.

When I was in college, I drove up this road whenever I had a free weekend and could convince a few friends to join me, and these thoughts resided on the periphery of my consciousness. They added a heaviness to the already solemn feeling of the trees, still massive even if second growth. We drove past it quickly, uttering righteous condemnations, and moved further into a wilderness that was remarkably accessible. After what seemed like hundreds of turns, the road ended at a small parking lot for about fifteen vehicles. There were log posts to hitch the mules, which the National Park Service occasionally used. It was a quick 2.5

mile hike up to Olympic Hot Springs, a campground shaded by old growth trees. Some forty years ago, a sizable resort with several buildings stood here adjacent to the springs, but all traces of those buildings have been removed. In its current form, the sulfur-scented water simply leaks from the hillsides into shallow pools, dammed up by rock and mud berms carefully sculpted by prior bathers. Having traveled a journey up from a mysterious and magma-scalding source, the water steams on the surface of the pools. Those pools highest on the hillsides were almost too hot to bear.

Returning thirty years later, excitement and expectation coursed through my veins. I was excited to return to a spot that held particular significance for me. My expectations of what a restoration effort on such a grand scale might make me feel were less clear. In some vague way, I hoped that seeing it might provide me with a much-needed transfusion of optimism. On a practical level, I was also a little anxious about having enough food for the two days I would need to complete the twenty-mile round trip. The weather, 75 degrees and sunny, prodded me on. I needed to hike in today and leave tomorrow for the return trip. I loaded my backpack and started down the road now only open to nonmotorized travel.

When I hit the washout point, the road predictably ended. But it's a termination that takes some time to process. I was used to seeing bridges in the process of repair or replacement, but seeing a road completely washed away, with nothing remaining of the detached section, created an odd sense of displacement. The old way is completely and incontrovertibly gone, a fact that the raging river makes clear. The new path, a footpath hacked into the forest, is marked by a small temporary sign. A mile in, knowing that I eventually needed to be on the other side, I began to wonder if the Park Service had installed a walking log bridge that crosses the river. There is no way to cross this river on foot. It wasn't clear from the temporary signs, and I hadn't found a posted map at the trailhead. It seemed unlikely that a seventy-foot-long log bridge would have been built. Did the road cross the bridge again? Seeing the wreckage of old campground shelters flattened from previous floods did little to calm my anxiety. I knew this place, or I thought I did, but the way I interacted with it was new. I would have to keep walking along at my modest pace through the forest primeval, uncertain that I would reach my destination. If unsuccessful, I wouldn't have time to return to the beginning and attempt it again. I had to get it right this time.

After a couple of miles, the temporary trail returned to the former road. I walked the deserted asphalt and passed similarly deserted Park Service buildings.

An hour later, my worries vanished as the road crossed the river over an elevated bridge that had not been removed. The road was covered in a thin layer of decomposed needles. In many places the asphalt had crumbled and the vegetation had crept inward, overtaking the narrow road shoulder. Eventually, two people on mountain bikes passed me, and the rusty chain squeak from the bikes delivered associations of utility workers pumping the handles of a handcar on their way to repair a section of broken-down railroad. I trudged on, hoping to witness a repair of another type, ecological in nature and running sinuously rather than in parallel lines toward the horizon.

After almost two hours of hiking, I reached the site of the former dam and walked the wide concrete path along the top of one of the remaining flanks of the dam. From this perch, I looked down two hundred feet to the river rushing through the canyon (figure 11.5). The current was fierce, and from this height

11.5 A view of Lake Mills, now drained because of the removal of the Glines Canyon Dam, and the Elwha River now free to flow downstream.

it was vertigo inducing. Although the rapids would certainly mean death to any human who tried to pass, I had read that they were possible for returning Chinook. Looking out toward the former lakebed, young thirty-foot-tall red alders and black cottonwood trees packed the entire area, the succession process kick-started with these fast-growing trees. Within the next couple of decades, they would thin out and allow the Douglas firs, Sitka spruce, western hemlocks, and red cedars to gain hold and eventually dominate the canopy. At the center of the valley, the stream carved out new braided channels in the loose gravel. The ponded silt and sand now washed downstream to Salish Sea, and the smaller gravel and boulders reclaimed the streambed. This material, mobilized by the erosive force of the stream, beckoned the keystone species to make the upstream journey.

The scene in front of me was powerful, and I reflected on the efforts of thousands of people over the past twenty-five years. I wondered if the river would be filled with returning salmon twenty-five years from now, similar to the video I'd seen on Castle's phone. Complete restoration of the river relies on a series of choices made by people all connected to the Elwha in their own ways. It would be easy to be satisfied with the massive accomplishment of removing of these dams. The river is free again, but the river is not fully restored.

Will we forgo convenience and allow a river to chew away at our roadways in its quest for the proper substrate for its bed? Will we exchange the easy abundance that hatcheries provide for something less bountiful in our immediate future? Will we commit to the original vision of restoration, something not yet achieved at this scale elsewhere? Will we rise to a level of restoration that forces each of us to change our relationship with the river and, in many ways, with each other? The flourishing vegetation in the former lakebed feeds my hopes for this ideal scenario. The river, searching for its form, confirms that this process is still evolving. As McHenry and Peters commented, the story is unfinished. Here, in this newly freed river, there is a chance to make nature whole again and in some way atone for the shortsighted and unjust decisions of the past. The river is doing its part; I remain hopeful that we will do ours as well.

CONCLUSION

RETURNING TO WATTS BRANCH

I wanted to return on two wheels. The bike path along the stream was one of many investments the District of Columbia had made along the Watts Branch corridor. Jim, the bike trail project manager for the transportation department, always asked me how the stream project was going. My answer, some variation of "just waiting on . . ." was always met with a knowing smile. There was no judgment in his unspoken response, but I certainly felt guilty that the stream project was delaying the installation of the bike path. As a fellow inhabitant of the bureaucracy, he understood, but these repeated conversations became too familiar to me. Who wants to be the one holding up a bike path? I shelved these earlier conversations, hopped on a bikeshare, and began the most uncertain of victory laps.

———— ◆ ————

Given my personal history with the area, it took me some time to absorb what was in front of me. I had spent so much time here—time when I often felt inefficient. It's been nearly a decade since the stream-restoration project was completed. My memory of the stream is tied to that period right after restoration when the banks were still raw. The vegetation, not yet established, doesn't provide the stability required. The boulder structures, fresh from the quarry, telegraph a sort of overconfidence. They've not been tested. The entire reworked landscape holds its breath, praying for the floods to hold off long enough to allow the stream to reestablish. Praying for it all to work.

I approach the first park node along Minnesota Avenue, where stormwater planters installed to clean runoff to the stream erupt with vegetation. This grassy space and outdoor pavilion behind the McDonald's is one of the "nodes" that make up Marvin Gaye Park. Not large enough for any official sports, it provided some shade, some seating, and enough room perhaps to toss a football. There are no long views or interesting topography that Frederick Law Olmsted turned into resonant landscapes. There is just the stream to unify these areas.

As part of the park rehabilitation, the parks department had installed benches and patios inscribed with lyrics from Marvin Gaye's song "Mercy, Mercy Me." The words of the native son who sang about the broader environmental crisis seem to capture the entirety of our environmental challenge. The song is the definition of the word *resonant*. It is astounding to think that a person who received mostly abuse from his neighbors during his childhood would end up writing a song that brought him adulation from across the globe. From that humble and now demolished row house a block from the banks of Watts Branch, Gaye's words float downstream in space and time to this bench where I sit fifty years later. The words, carved in granite, are not marred by any graffiti.

I pass a middle-aged Black woman walking her tiny dog through the seating area. The dog keeps my reverie in check with a shrill greeting. I make an easy joke about how the park is all his. The woman laughs and continues on her unhurried walk. I intentionally stop here because the stream in this section makes a near 90-degree turn before it completes its final run to the Anacostia River. This is also one of the locations where, due to work done sometime in the past one hundred years, the stream was confined to the shape of a paved trapezoid. Although water flowed through here, it was not a stream.

In addition to nearly two miles of stream bank grading, the Watts Branch restoration project took this hardened section out. I had waited nearly eight years to see it come out, and when it did, I captured the clunky movement of the machines as if it was my child's first dance performance. It was exhilarating. So it is a bit disheartening when I see that same section has been paved over again. In the ever-evolving trajectory of stream bank work, this section that had been "restored" at great cost has been replaced with sandbags that seem to have been filled with cement rather than sand. The entire bulwark is covered with white concrete (figure c.1). It has all the signs of an emergency repair. This bank isn't going anywhere now. The flashy deluges originating in Maryland haven't changed, and this was probably the best option to prevent a major blowout of the stream and the park. I feel my hardened cynicism returning.

I bike past a woman pushing her child in a stroller who smiles as I slow down. I am keenly aware that the vegetation had grown, and in some places branches

c.1 An upstream view of a section of Watts Branch that had been restored but is now restabilized.

hang over the path as if trying to reach out to the human passersby. I want to talk but hesitate, knowing that this area was plagued by violence when I worked here. The violence affected people I knew, and I assume that stopping might imply danger, and I don't want to cause unnecessary alarm. In addition to providing a buffer that would stabilize the stream, this vegetation provides ample locations to lie in wait or to drag a victim. Feeling responsible for this tangled mass of trees and shrubs, I've never felt so conflicted about a riparian buffer. The woman, unconcerned by my internal deliberations, happily continues on her afternoon stroll.

Around the corner, the long-established Lederer Youth Gardens sits just fifty feet from the stream. No one is present this afternoon, but the crops indicate that significant labor has been invested here. The full beds promise an ample harvest. Collards, tomatoes, lettuce, and other crops flourish behind the tall fences installed to prevent vandalism. On the side facing the street, paintings of beekeepers, cherries, cabbage, and strawberries overlook the wide roadside shoulder where a long depressed grassy basin sits ready to capture stormwater from the street. Hand-painted signs notify residents of the time for the free veg-etable give away, and a longtime district employee's email address is listed for more information.

Along the trail I spot several signs that describe some natural feature. In one spot, a small prairie had been planted. The solid mass of three-foot-high peren-nial vegetation readies itself for local pollinators. Signs provide information about different trees and how the wire cages circling the trunks are necessary to prevent damage from the beavers that have returned to the stream. In some places twenty-inch-diameter trees have been protected. The beaver population must be healthy.

At several points I walk down to the stream to get a better look at the structures. The sturdy cross-vanes have held up even though a couple of granite boulders have rolled into the pools in a few locations. In one location, approximately halfway along the district's section of the stream, a floating trash trap has been installed. Although the sign has fallen, the metal cables that anchor the contraption to the banks are taut and secure. The ubiquitous single-use plastic bottles float inside the trap, seemingly struggling to find meaning in their ten minutes of unremarkable convenience that ends in this stream. Restoration is not a point in time but a pro-cess that requires personal commitment as much as proper design.

At the center of the park where the drug dealing had once been the most obvious, a splash pad and playground disarms the landscape. Some dads talk together while their kids use the equipment. Across the street that runs through this node, a couple of large tents have been pitched. Several seated men are play-ing a card game and others are just talking. The purpose of the tents seems to be for shade rather than use as a homeless encampment. Although not a desired activity in the minds of staff, the park is being used in a way that doesn't violate any significant laws.

Moving eastward toward the district line, I pass a newer, four-story apartment building elevated six feet above the surrounding floodplain on concrete pillars. Apparently they had not encountered the challenges I had with the district per-mitting process. Four levels of sliding glass doors look out toward the stream, but no porches were included. It has all the signs of a development shaped by shrewd attention to profit margins, but all the units appeared to be vacant.

Up at the district line, a new recreation center occupies what had been a vacant lot on the north side of the stream. Clad on its front side with a decorative aluminum facade and painted trendy modern gray, the new center stands out when compared to the older small brick one that still exists on the southern side of the stream. I don't notice anyone at the new facility, but several people are gathered in the playgrounds that surrounded the older one. A few younger men play dice on the sidewalk not far from a new statue of Marvin Gaye.

I decide to turn back, still feeling uncertain about the state of the stream and the overall neighborhood. What did all of this investment provide for this neighborhood? What was the significance of this stream project?

On my way back I encounter an older Black man wearing a blue bucket hat who is trimming his white picket fence line with a string trimmer. His yard is tidy and faces one of the thin sections of park that is only wide enough for the stream. I approach him, and he stops his work. Jogo, as he identifies himself, is a friendly older man with a warm African accent. We begin a conversation, and I learn that he has lived in his house for thirty-eight years. He remembers the work on the stream even though the section across from his house didn't require earthmoving or in-stream structures (figure c.2).

c.2 Looking upstream at a restored section of Watts Branch with numerous cross-vane structures.

I ask him what he thinks about the stream now and if it has improved. He speaks of improvements in the neighborhood and gestures toward some "nice new Spanish neighbors." He points at the nearby corner and says it used to be a hot spot for drug dealing. Things are better, he says, and I don't get the sense that he is trying to provide some assurance that I may appear to be needing. He tells me that the pandemic has worsened some things—how people are cooped up—and some random shootings happen that still make no sense. It is not an answer to the question I asked, but perhaps it is reflective of what this stream meant. He was telling me that the health of the stream is the health of his community (figure c.3). As we look toward the stream, I notice several river birches that were planted as part of the project. They are now at least thirty feet tall and shade the stream, cooling the water that fell as rainfall on scorching parking lots in Maryland. The trail at this point runs along Jogo's street before it crosses over

c.3 Jogo tidying up his yard.

the stream. Jogo tells me he advocated for that new bridge. "Lots of people use it, bikers, people exercising," he says with obvious approval.

This stream is just one thread in the fabric that is the larger community. The fabric is textured, dense, and brightly colored. Yet it strikes me that this one thread might hold everything else together. When this thread loosens, these varied efforts fail to unify into something more than a random assemblage of interventions by an evolving group of well-intentioned people. The pattern is lost and directionless without the current. When taut, the thread pulls these actions together and responds to Marvin Gaye's mellifluous call to action. The stream, responsive to our collective actions, tells us how we're doing.

I jump on my bike and head back. I slow down as I approach a middle-aged Black man walking in the opposite direction who has stopped in the middle of the path. He's closely inspecting the fruit of a mulberry tree that hangs over the trail. I have also enjoyed these berries on many occasions. We exchange head nods, and he continues on. It seems that the berries aren't quite ripe, but they're almost there.

EPILOGUE

I n a conference full of participants enthusiastic about beaver dams, no one looks askance at two young women at a table full of stickers and banners advocating against dams. The quilt hanging from the table is made from a patchwork of smaller fabric pieces etched with hand drawn messages: "Imagine a dam-free future" and "Save Idaho Salmon." The dams Lilly Wilson and Lizzie Duke-Moe are protesting are not the ecologically beneficial impoundments of a keystone species but the four massive hydroelectric dams on the lower Snake River. These four dams, operated by the Army Corps of Engineers, produce 993 megawatts of power annually, which represents 4 percent of the region's energy use.[1] They also have led to the precipitous decline of salmon runs up the Snake River. The Snake River watershed drains the majority of Idaho, wide swaths of eastern Washington and Oregon, and relatively smaller, yet still sizable slivers of Wyoming, Nevada, and Utah. For Idaho, a state full of mountain streams that have historically teemed with fish, the state motto of *Esto perpetua* doesn't apply to the salmon that are on a path to extinction.

The fish viewing room could be the world's saddest aquarium. A large carp swims alongside a couple of confused looking smallmouth bass that peered at me through the algae covered glass. A lamprey or sucker of some sort hangs on to the yellow-edged bottom corner of one pane. I assume it is alive but only because it is attached. Prominently displayed is the "Fish Counting Station," a readout of yesterday's and year-to-date tallies of fish that have passed through the ladder. The red digital numbers emanated from the obsidian display are reminiscent of an earlier generation digital alarm clock or a command center from an eighties sci-fi movie. As of July 24, American shad were the big winners with 187,513 passing year-to-date, Chinook 84,685, steelhead 4,218, sockeye 1,794, and Coho zero. Prior to the dams on this river two to three million Chinook and up to eight million salmon regularly passed through this portion of the river.[2]

I'm in the belly of the lowest of four dams on the Snake River, 1,000 miles downstream from the pristine mountain streams of the Salmon River headwaters. The salmon runs of the Snake River are the longest runs with the highest elevation gain on the planet. The fish gain 6,500 feet in elevation on their trip from the ocean to the headwaters of Pole Creek, the highest point in their migration. Although biologically equipped to survive this incredible journey, we have thrown these fish challenges they are not equipped to handle. If they make it up the fish barriers, migration of smolts back to the ocean is even more precarious. Not represented on the production tally is the species on the model next to me, a river sturgeon about five feet long. "That one is getting close to the size you can harvest," a friendly voice tells me. "We get them up to about ten feet here in the Columbia and lower Snake." The helpful voice belongs to Gary Gardner, an Army Corps of Engineers volunteer at the Ice Harbor Dam Visitor Center.

Gardner wears a crumpled red-and-white hat that displays the Corps' castle logo. Despite possessing a substantial physical presence, he gives off an innocuous vibe. He wears a comfortable shirt and shorts, which makes sense given that it's nearly 100 degrees outside of this bunker we are currently inhabiting. Gardner and his wife, now retired, have volunteered for the past three years as interpreters at this visitor center housed in the lowest of the four Snake River dams. The Corps provides a free campsite for them at the nearby Charbonneau Park Campground. This modest perk comes in handy because he and his wife recently sold their house and bought a camper. He has lived in the tri-cities area of Washington most of his life and is happy to be spending his retirement next to the clear waters of the reservoir. Although it wasn't a transaction in the strict sense, the Army Corps is getting a phenomenal deal by having Gardner carry its proverbial water.

I tell Gardner that I'm writing a book on river restoration and have spoken with some younger advocates who are pushing to have the lower Snake River dams removed. "They call themselves the Youth Salmon Protectors," I mumble sheepishly. I'm not sure what to expect because I had to pass security to get to this point. "Oh, a group of those folks descended upon us a couple of weeks ago. Very nice, very polite, but they had their agenda," Gardner says, reassuring me that he is not offended by the purpose of my visit. Gardner explains that he had worked in two industries that might also have attracted some attention from this group. 'I worked in the pulp and paper industry for twenty years, then for the nuclear industry for another ten. Went from one devil to the next," he jokes. Despite his verbal parries, I sense that I'm dealing with a decent guy. Gardner is the kind of guy you wouldn't mind chatting with at a cookout or chewing over the latest sports headlines. I ask for his perception of the group.

"They were polite, but some, like the lawyerly thing, asked questions they already knew the answer to," Gardner says. Someone asked whether the 603 megawatts of production from the six turbines in the dam was per year or per day. "I hadn't really thought about it," he tells me. He described the barging operation the Corps operates that bring smolts downriver all the way through all eight dams on the lower Snake and Columbia system (figure e.1). It turns out

E.1 A discharge pipe for smolt passing down through Ice Harbor Dam.

that migration downstream by the smolts is the biggest challenge to sustaining salmon runs on these rivers. "I mentioned that the fish were collected from the wild," Gardner says, "but it turns out there was a fisheries biologist in the group. They said that's not how it happens." Gardner seems more annoyed at having provided bad information than at being corrected. He likes interacting with the public but is not fond of giving lectures to a highly skeptical crowd. More than thirty people were hanging on his every word, he tells me. "If they want to beat up on a sixty-five-year-old fat guy, fine then."

Gardner is more knowledgeable than the way he presents himself. When I ask whether he thinks the dams will ever come out, he lists the full range of issues that make such a huge decision difficult. He's also honest about impact. "The Corps, they tell me 'don't lie,' the dams do impact the fish. But the thing people don't realize is that the Corps is spending billions to address the situation. I think the advocates are going after the wrong thing. Go after the long liners who have mile-long lines and are taking all of the salmon out of the ocean. Go after the foreign boats that come into our fishing waters two miles out. They think that by removing the dams the salmon will recover dramatically—but I think they're wrong."

Gardner continues, happy to have the chance to rebut this group now that it's a fair fight. "They don't think that flooding will be an issue. There's 260,000 people living in the tri-cities now. You can't just move all of them somewhere," he says, highlighting a benefit of the dams few recognize. "The problem is that they don't live here so they don't think about that." He mentions the orchards and extensive agriculture industry that is supported by the reservoirs these dams create as well as the large shipments of wheat that are efficiently and cheaply delivered from barges upriver in eastern Washington. The war in Ukraine is creating global shortages of wheat right now, and starvation is being predicted for many people in the developing world. Gardner presents a frank and sobering assessment.

I ask if local folks are talking about the potential removal. "You see there are actually two Washington states," Gardner says. "The three counties west of the Cascades determine everything that happens in this state. We send our people over there, but they come back with nothing." This may be true, but Gardner's astute political assessment misses one thing. Those advocating to remove the dams are coming from the east and out of state as well. From the conservative bastion of Idaho, politicians and advocates are putting forward serious proposals to remove the dams along with plans to compensate for the benefits the dams have provided. Gardner may have more water to carry for the Corps.

Lilly Wilson, part of the Youth Salmon Protectors (YSP) group that had grilled Gardner, told me she was surprised by the scale of the dam. She knew it would be big, but the angular behemoth dominated all in this semiarid landscape where the rounded hills appeared smooth and empty in the distance. She's been involved for only a year but has already learned the lesson of how to frame the message. "We're in deep red Idaho. We can't say sustainability. We have to talk about conservation and how removing the dams could help hunters and fishers."

Wilson invites some of these hunters and fishers to come their events, and a few from a recent fisherman's expo came along on their trip to the dam. "It was exciting to watch these guys over fifty learn along with the larger group of teenagers." "They love the stickers, and it gets them talking about their experiences fishing for salmon. You can tell how much those experiences mean to them." But she has learned that there are touchy issues. "They always talk about the lost energy. They need a little convincing on the potential for solar power." This nay-saying doesn't seem to affect her. I mention that Governor Insley and Senator Murray from Washington have come out with a framework in response to Idaho Representative Mike Simpson's and ask if this is positive. Wilson is optimistic. Things are moving. It's only a matter of time.

Wilson points out that this issue has been raised before. For an art project in elementary school, she remembers looking through a 1982 *National Geographic* article that explained how the dams were killing the salmon. But it's as if time is on her side. The naysayers talking about lost energy production will soon move on. The future without dams is on the horizon. She's off to college at Boise State where she plans to start a student chapter of YSP as she studies ecology and plant science. Her advocacy for the dams has already provided her with an education in political science; I think she could make some progress with Gardner.

As I drive through the agricultural area surrounding the tri-cities, I'm struck by both the potential and the waste that is evident. People are watering their three-acre lawns with farm-sized sprinklers at 11 a.m. when it is already 95 degrees and is predicted to top 100. Someone is watering the grass on a small horse or cattle grazing area, but no livestock are present to enjoy the shower.

In terms of potential, the area receives more than 300 days of sunlight a year, and space is abundant.[3] Despite the presence of many orchards and some small ranches, the dry, empty land eclipses the area in production. Some plants and animals survive in this arid land, but it's undeniable that empty space is abundant. Gardner told me that one turbine provides enough energy to support approximately 120 households for a year. I wonder how many acres of solar power would be needed to provide the equivalent amount of energy. Whatever that area is, there is enough of it in central Washington. Along the ridges are scattered wind farms. Spaced far apart, they seem to beckon for company. I've rarely been in a place where wind and solar not only seem possible but in many ways are the obvious solution.

I'm on my way to the confluence of the Snake and the Columbia rivers to visit a state park named after Sacajawea. I pull onto the shoulder of the road to type in some notes from my recent conversation with Gardner and am surprised to hear rain hitting the roof. I'm receiving an impromptu car wash from a sprinkler set in a field of potatoes. The sprinkler shoots Snake River water onto the blazing hot asphalt of Highway 124 where it immediately evaporates into the atmosphere.

When I get to the park, I survey the grassy landscape dotted with shade trees. The water surrounding me is flat except for the wake kicked up by a passing fishing boat and a jet ski. A few families wander around and head toward a small museum that details the Lewis and Clark expedition. I speak to the ranger who is taking a break outside. She tells me it's hard to know that you're at the confluence of two rivers given the flat water. The two rivers have formed a giant lake that surrounds this peninsula of manicured turf. A pelican lazily flies away, apparently comfortable in this landscape reminiscent of a golf course in Florida.

Once I get my bearings, I look up the Snake River where a couple of iron drawbridges frame the scene. Their form further constrains the perspective of this river, as if it is being divided and sectioned for our purposes. Despite this emasculation of its fundamental character, many people are enjoying the area. Speed boats whip by on the calm water. Fully rigged fishing boats hunt for whatever fish swim underneath, desperately searching for a current to direct them. The local economy is supported by this activity. Tow packages have been selected at the local Ford dealership to accommodate these rigs. Depth finders have been purchased to outfit a fully modern fishing boat. And all manner of boat toys ranging from water skis to inner tubes have been purchased and packed to ensure that we extract a full day's enjoyment out of this water body.

Lizzie Duke-Moe got involved about three years ago. She volunteers for the YSP, which is sponsored by the Idaho Conservation League. She's been helping with banner drops on bridges that cross the Snake River. "So simple, but it generates a lot of press, and people start asking questions," she says. Her grandfather was born in Norway, and she has tapped into a deep-seated genetic appreciation for salmon. For backup, she's thrown herself into reading a philosophical tome, *Being Salmon, Being Human*. "Hardest book I've ever read", she tells me. It is a recommendation from someone who doesn't expect things to be easy.

Duke-Moe has met most of the key representatives in Idaho even though she hasn't finished high school yet. She met Congressman Simpson, who she described as a "happy guy who wants to be outside." The fact that she is seventeen and he is seventy-one is only the most obvious of their differences. She was raised by lesbian moms, and he had a solidly conservative Christian upbringing. During their meeting, she and her colleagues asked the congressman to speak about how the salmon mattered to him. He spoke in "terms of God and our responsibility to God's creation." "He has a wife, and I have a girlfriend, but we decided, despite all of that, we should save the salmon."

This disarming anecdote of polar opposites coming together belies something more significant in the works. In spite of the potential blowback, Congressman Simpson, from the deepest red state in the country, has released a proposal to remove these four dams. Advocacy has come from many angles, and the YSP would be first to acknowledge that their efforts are just a part of the larger effort. But their modesty doesn't temper their faith in a future without the Snake River dams. "I know that I will see a Snake River without dams in my lifetime," Duke-Moe tells me.

Although not laying out the specifics, Simpson's framework is decidedly broad and comprehensive. Removing the Snake River dams affects the transportation route for major agricultural shipments of wheat and other products. This transportation is by far the cheapest and most predictable mode. Replacing it with rail is possible with the expansion of rail routes, and this is included in the plan. But preventing cost gouging that might come from a rail industry that has no competition is also a concern. The plan allocates funds to ensure that shipping costs don't exceed preremoval levels. For all of these shipping and agricultural impacts, the plan allocates a tidy $5 billion.[4]

Replacement of the energy generated by the dams is another major component. Millions of solar panels would need to be installed to compensate for the loss, and $16 billion is proposed to pay for this transition.[5] The area is abundant in sunshine, and the stimulus to the renewable energy business could create thousands of jobs.

The altered shorelines of the towns and cities along the Snake River would also need to be addressed. A less constrained river would result in a more hydrologically dynamic river that may require reinforced areas as well as flood-plains that could absorb flood waters. The framework includes a round figure of $2.3 billion for this work.[6]

Duke-Moe recalls her trip to the dam with the group. She was struck that no information provided there recognized the massive dissipation of salmon runs that followed construction of the dams. Before the dams were installed, Chinook runs of 200,000 to 300,000 occurred every spring.[7] "The signs said the biggest problem were the small irrigation dams," she tells me incredulously. Talk about the pot calling the kettle black. Duke-Moe is heading to college in the fall. Her application letter to Brown University spoke of her meeting with Congressman Simpson, and the application committee was understandably impressed. Before she heads east, she'll be attending a leadership training course through the Earth Guardians in New Mexico. Although she's excited about college, Brown doesn't have any classes on salmon. She's determined to find some way to expand her knowledge and retain that connection to her home and the river that needs her. I trust she will find a way.

—•—

The placid water presents other challenges. David Cannamela is a retired Idaho state fish biologist who accompanied the YSP and corrected Gary Gardner's fish facts. He explains that the warmer water allows predator fish to thrive. Bass, walleye, and pike minnow balloon to unnatural numbers in the lakes and feast on salmon smolts. "There are two parts to the impacts of the dams," he tells me.

"Most people know about the smolts getting chopped up in the turbines or getting the bends as they experience massive pressure changes. That does happen. But the other major part is the delay and disruption in the migration of the smolts to the ocean."

"The real issue is that the fish are on a time line," Cannamela continues. When the fish return to the ocean, numerous physiological changes are occurring: skin function, kidney function, and the ability to osmo-regulate in salt water. The smolts make these changes on a biologically regulated time line, but their path to the ocean has been significantly delayed by the still waters. "When you've disrupted this very precise timing, you've taken away the ability of the fish to survive in the new environment." Termed *delayed mortality*, this river-scale disruption has led to a smolt to adult return (SAR) rate of less than 1 percent in the Snake River. A 4 percent SAR is required for healthy populations that can be sustainably harvested.[8]

The Corps touts the number of smolts that pass through the dam, and Cannamela relates this situation to an ambulance ride after a car accident. Although the patient in the ambulance is technically alive, the condition of the patient in the hospital is another matter. He rails against those who point to problems in the ocean. "If the ocean was the problem, why do the lower tribs in the lower Columbia have a higher smolt to survival rate and still see relatively strong returns?" Before the dams were installed, the runs up the Snake River were as strong or stronger than those in the closer Yakima and John Day rivers.

I ask Cannamela what it's been like working with the YSP. He smiles, searches for words, and tells me that it's been energizing. He wonders where they get their optimism having only seen a river that is on life support. He has been seasoned by twenty-eight years in government but still seems up for the fight. "It's going to happen," Cannamela summarizes, "but it's not happening as fast as it should." "We're standing at the crossroads of what would be the largest river restoration ever undertaken. We can restore the river and all the river does," Cannamela says, getting lost in a list of cascading ecological and biological benefits. "I tell people river restoration is the goal, dam removal is the tool."

The enormous ballpark figures in Simpson's framework can both stupefy and disempower the casual observer; $33 billion is a lot of money. Regardless of the price tag, we spend whatever is necessary when we find that it is necessary: $2.2 trillion for the CARES Act to provide enhanced unemployment benefits and debt relief for businesses; $1.94 trillion annually for the Department of Defense; $17.4 billion to bail out the auto companies; and nearly $4.5 billion annually on corn subsidies alone.[9] What price are we willing to pay to have salmon in the Snake River basin? This system isn't just four Elwha rivers. It's an entire river system that occupies portions of six states. How much do we value a river system that functions like a river?

In this semiarid valley blanketed with water delivery systems that support vineyards and orchards of apples as tempting as those in Eden, it's difficult to imagine an undammed river that could still provide water to our farmers. Currently embodied as a series of still pools that allow for the cheap transport of goods, it's hard to imagine that a river with rapids might provide useful transportation. It takes a leap of faith to believe that we can remain safe next to a river that hasn't been controlled. It takes a leap of faith to imagine that we can still benefit from a river that may flood us.

Wendell Berry says that we must acknowledge the essential nature of rivers:

> To a river, as to any unfettered natural force, an obstruction is merely an opportunity. For the river's nature is to flow; it is not just spatial in dimension, but temporal as well. All things must yield to the impulse of water in time, if not today then tomorrow or in a thousand years. If its way is obstructed then it goes around the obstruction or under it or over it and, flowing past it, wears it away. Men may dam it and say that they have made a lake, but it will still be a river. It will keep its nature and bide its time, like a caged animal alert for the slightest opening. In time it will have its way; the dam like the ancient cliffs will be carried away piecemeal in the currents.[10]

Berry's expansive and optimistic time line is invigorating as we sit in the fetid, stagnant backwaters of the collective impacts humans have had on our rivers. His rendering of the primordial spirit of the river hits the mark, but it misses what rivers need at this moment—an expansion of human nature. I return to my first experience with the Elwha dams. Thankfully some people imagined that the river could be more than a source of water and energy. This imagining is the first step in any restoration: imagining a future in which the rivers we love can function as rivers. As David Cannamela deftly concluded, "The fish need a river. That's what it comes down to."

NOTES

1. FUMBLING FOR BANKFULL: DAVE ROSGEN AND THE STRONG CURRENTS OF A STREAM-RESTORATION METHODOLOGY

1. UMass Amherst Riversmart Communities, https://extension.umass.edu/riversmart/resources/what-fluvial-geomorphology-fgm.

2. Andrew Simon, Martin Doyle, Matthias Kondolf, et al., "Critical Evaluation of How the Rosgen Classification and Associated 'Natural Channel Design' Methods Fail to Integrate and Quantify Fluvial Processes and Channel Response," *Journal of the American Water Resources Association* 43, no. 5 (October 2007): 1125.

3. Rebecca Lave, Martin Doyle, and Morgan Robertson, "Privatizing Stream Restoration in the U.S.," *Social Studies of Science* 40, no. 5 (September 2010): 682; originally cited in E. S. Bernhardt and Margret Palmer, "Synthesizing U.S. River Restoration Efforts," *Science* 208 (April 2005): 637.

4. Bernhardt and Palmer, "Synthesizing U.S. River Restoration Efforts," 637.

5. Lave, Doyle, and Robertson, "Privatizing Stream Restoration in the U.S.," 686.

6. Lave, Doyle, and Robertson, "Privatizing Stream Restoration in the U.S.," 689.

2. THE BOG ARCHITECT: REIMAGINING STREAMS AND STORMWATER ON THE COASTAL PLAIN OF THE CHESAPEAKE

1. Aimlee Laderman, "The Ecology of Atlantic White Cedar Wetlands: A Community Profile," *U.S. Fish and Wildlife Service Biological Reports* 85 (July 2021): iii.

2. Reed F. Noss, Edward T. LaRoe III, and J. Michel Scott, "Endangered Ecosystems in the United States: A Preliminary Assessment of Loss and Degradation," *National Biological Services Biological Report* 28 (February 1995): 58.

3. William J. Mitsch and James G. Gosselink, *Wetlands*, 3rd ed. (New York: Wiley, 2000), 473.

4. Laderman, "The Ecology of Atlantic White Cedar Wetlands," 29.

5. Karen A. Terwilliger, "Breeding Birds of Two Atlantic White-Cedar Stands in the Great Dismal Swamp," in *Atlantic White Cedar Wetlands*, ed. Aimlee D. Laderman (Boulder, CO: Westview, 1987), 215–17.

6. Emily S. Bernhardt, Margaret A. Palmer, and J. David Allan, "Synthesizing U.S. River Restoration Efforts," *Science* 308, no. 5722 (May 2005): 636–37.

7. Solange Filoso and Michael Williams, "Sources of Iron (Fe) and Factors Regulating the Development of Flocculate from Fe-Oxidizing Bacteria in Regenerative Streamwater Conveyance Structures," *Ecological Engineering* 95 (2016): 723–37.

3. LEGACY SEDIMENT: DOROTHY MERRITTS AND ROBERT WALTER DIG BACK IN TIME IN LANCASTER COUNTY, PENNSYLVANIA

1. Chesapeake Bay Program (EPA), "Conowingo Dam," https://www.chesapeakebay.net /issues/threats-to-the-bay/conowingo-dam.
2. Jeff Hartranft (ecologist, Pennsylvania Department of Environmental Protection), in discussion with the author November 2022.
3. Tom Schueler and Bill Stack, "Recommendations of the Expert Panel to Define Removal Rates for Individual Stream Restoration Projects," January 2014, https://d38c6ppuviqmfp .cloudfront.net/documents/Final_CBP_Approved_Stream_Restoration_Panel_report _LONG_with_appendices_A-G_02062014.pdf.
4. Michael J. Langland, Joseph W. Duris, Tammy M. Zimmerman, and Jeffrey J. Chaplin, "Effects of Legacy Sediment Removal on Nutrients and Sediment in Big Spring Run, Lancaster County, Pennsylvania, 2009–15," *Scientific Investigations Report* 5031 (2020): 18, https://pubs.usgs.gov/sir/2020/5031/sir20205031.pdf.
5. Mike Rhanis (senior GIS specialist, Water Science Institute), in discussion with the author November 2022.

4. THE HUMAN BEAVER: MEGA- TO MICRO-ENGINEERING SOLUTIONS IN GREATER CINCINNATI

1. Robert J. Hawley, "Making Stream Restoration More Sustainable: A Geomorphically, Ecologically, and Socioeconomically Principled Approach to Bridge the Practice with the Science," *BioScience* 68, no. 7 (July 2018): 7, 517–28, https://academic.oup.com /bioscience/article/68/7/517/5034097.
2. Dan Horn, "West Side Neighborhood's Luck Running Dry Again," *Enquirer*, September 2, 2015, https://www.cincinnati.com/story/news/2015/09/01/south-fairmounts-luck-running -dry/71520244/.
3. Gray and Pape, Inc., "Work Plan for Additional Archaeological Investigations and Support for Mitigation, Lick Run VCS," May 2017: 5.
4. Strand and Associates, "Lick Run Comprehensive Design," Report for Metropolitan Sewer District of Greater Cincinnati, Ohio, March 2012, https://archive.epa.gov/r5water /lowermillcreek/web/pdf/lickrun-comprehensive_design_201203.pdf.
5. Robert Hawley (principal scientist), in discussion with the author November 14, 2022.
6. Lick Run Greenway, Metropolitan Sewer District of Greater Cincinnati, "Project Groundwork," https://www.projectgroundwork.org/projects/lowermillcreek/sustainable /lickrun/.
7. Paul Muller (director, Cincinnati Preservation Association), in discussion with the author January 2022.

5. BEAVER WRANGLERS: FACILITATING FUNCTIONAL RIVER RESTORATION IN WESTERN WASHINGTON

1. Emily Fairfax and Andrew Whittle, "Smokey the Beaver: Beaver-Dammed Riparian Corridors Stay Green During Wildfire Throughout the Western United States," *Ecological Applications* 30, no. 8 (September 2020): e02225, https://doi.org/10.1002/eap.2225.
2. Ellen Wohl, Anna E. Marshall, Julianne Scamardo, Daniel White, and Ryan R. Morrison, "Biogeomorphic Influences on River Corridor Resilience to Wildfire Disturbances in a Mountain Stream of the Southern Rockies, USA," *Science of the Total Environment* 820 (May 2022): 2, https://doi.org/10.1016/j.scitotenv.2022.153321.
3. Wohl, Marshall, Scamardo, White, and Morrison, "Biogeomorphic Influences on River Corridor Resilience," 9.
4. Michael Pollock, Timothy Beechie, Joseph M. Wheaton, Chris Jordan, Nick Bouwes, Nick Weber, and Carol J. Volk, "Using Beaver Dams to Restore Incised Stream Ecosystems," *Bioscience* 64, no. 4 (April 2014): 279–90, at 285.
5. "Explore: Middle Fork Snoqualmie Valley," Mountains to Sound Greenway Trust, https://mtsgreenway.org/explore/middle-fork/.

6. WISCONSIN TROUT: RESTORING DRIFTLESS AREA STREAMS AND MITIGATING FOR EFFECTS OF CLIMATE CHANGE

1. John F. Lyons, J. S. Stewart, and Matthew Mitro, "Predicted Effects of Climate Warming on the Distribution of 50 Stream Fishes in Wisconsin, USA," *Journal of Fish Biology* 77, no. 8 (2010), 1867–98.
2. Lee Bergquist, "Little Plover River Near Stevens Point Named One of America's Most Endangered Streams," *Milwaukee Journal Sentinel*, April 17, 2013.
3. City of River Falls, Wisconsin, "Storm Water Ordinance," n.d., https://www.rfcity.org/262/Storm-Water-Ordinance.
4. City of River Falls, Wisconsin, "Sterling Ponds Park Plan 2020," February 25, 2020, 12, https://www.rfcity.org/DocumentCenter/View/3926/Sterling-Ponds-Park-Plan-Final-Draft-?bidId=.
5. City of River Falls, "North Kinnickinnic River Monitoring Project," 2014 Report, 8, https://www.kiaptuwish.org/resources/.
6. "Wisconsin Groundwater Coordinating Council Report to the Legislature, 2022," 2, https://dnr.wisconsin.gov/sites/default/files/topic/Groundwater/GCCGWQuality/Nitrate.pdf.

7. RIVER CANE DREAMS: A PLANT THAT RESTORES CONNECTIONS

1. Leader News Staff, "Timber Thief Gets 30 Months in Prison for Illegal Logging, Forest Fire," *The Leader*, September 28, 2020, https://www.ptleader.com/stories/timber-thief-gets-30-months-in-prison-for-illegal-logging-forest-fire,71428.
2. Theodore Roosevelt, "In the Louisiana Canebrakes," *Scribner's Magazine* 43, no. 1 (1908): 47–60, https://sites.rootsweb.com/~lamadiso/articles/louisianacanebrakes.htm.

3. Jason Meador (biologist), in discussion with the author February 14, 2022.

4. Kelly Williams (In-lieu Fee Program Coordinator, North Carolina Department of Environmental Quality), in discussion with the author January 2023.

5. Robin Wall Kimmerer, *Braiding Sweetgrass: Indigenous Wisdom, Scientific Knowledge, and the Teachings of Plants* (Minneapolis, MN: Milkweed, 2015), 157.

9. COMMUNITY BONDS: ORGANIZATION AND COLLABORATION IN WEST ATLANTA

1. Bill Eisenhauer (community activist), in discussion with the author May 2023.

2. ParkPride.org, "Proctor Creek North Avenue Watershed Basin: A green Infrastructure Vision," 2010, www.parkpride.org/wp-content/uploads/2016/09/2010_pna_overview-1.pdf.

10. SOUTH RIVER ACTION HERO: ENVIRONMENTAL JUSTICE AND ACTIVISM IN SUBURBAN ATLANTA

1. Niche.com, "Johns Creek High School," accessed November 10, 2022, https://www.niche.com/k12/johns-creek-high-school-johns-creek-ga/.

2. South River Watershed Alliance, South River Forest Coalition, Margaret S. Brady, Allen P. Doyle, Joel Finegold, Joseph S. Peery, and John and Jane Does (Plantiffs) v. Dekalb County, Georgia, by and through its Board of Commissioners, and Blackhall Real Estate Phase II, LLC (Defendants), Civic File Action No. 21CV1931, Filed by Joel Finegold, February 12, 2021, https://www.stoptheswap.org/_files/ugd/773705_e2f0b0d8dd064a41a06776b7ebcf1f5f.pdf.

3. Kendall Glynn, "Cop City Explained: A Look at the Ongoing Controversy Surrounding Police Training Center," Decaturish.com, September 1, 2022, https://decaturish.com/2022/09/cop-city-explained-a-look-at-the-ongoing-controversy-surrounding-police-training-center/.

4. South River Forest Coalition, "South River Forest: An Overview," accessed November 12, 2022, https://www.southriverforest.org.

5. EPA, Office of Inspector General, "Report: Atlanta Is Largely in Compliance with Its Combined Sewer Overflow Consent Decree, but Has Not Yet Met All Requirements," May 30, 2018, https://www.epa.gov/office-inspector-general/report-atlanta-largely-compliance-its-combined-sewer-overflow-consent.

6. "Fifty Percent of U.S. Waterways Impaired by Pollution," *Smart Water Magazine*, March 21, 2022, https://smartwatermagazine.com/news/environmental-integrity-project/fifty-percent-us-waterways-impaired-pollution.

11. DAM REMOVAL AND COMPLICATED HISTORIES: UNFINISHED BUSINESS ON THE ELWHA RIVER IN WASHINGTON STATE

1. Mike McHenry (fish habitat manager, Lower Elwha Klallam Tribe), in discussion with the author April 2023.

2. National Park Service, Olympic National Park, Washington, "Elwha River Restoration Frequently Asked Questions," accessed May 2, 2023, https://www.nps.gov/olym/learn/nature/elwha-faq.htm.

3. McHenry, in discussion with the author April 2023.

4. Mike McHenry, et al., "2022 Elwha River Smolt Enumeration Study," Submitted to Olympic National Park by Lower Elwha Klallam Tribe, March 20, 2023.

5. Edward Abbey, *Desert Solitaire: A Season in the Wilderness* (New York; Ballantine, 1971), 50.

6. Abbey, *Desert Solitaire*, 59.

EPILOGUE

1. Eli Francovich, "Dam Power: Snake River Dams Are Not Big Power Producers but Play an Important Regional Role," *Spokesman Review*, March 8, 2021; NW Energy Coalition, "The Lower Snake River Dams Power Replacement Study," April 2018, https://nwenergy.org /wp-content/uploads/2018/04/LSRDS-study-4-page-overview.pdf.

2. David Cannamela (fish biologist), in discussion with the author October 13, 2022.

3. Tri-City Development Council, "Climate & Geography," n.d., https://www.tridec.org /climate-geography/.

4. Idaho Conservation League, "Columbia Basin Initiative: Congressman Mike Simpson's Proposal," April 2021, https://www.idahoconservation.org/wp-content/uploads/2021/04 /Simpson-Proposal-Overall-1.pdf.

5. Idaho Conservation League, "Columbia Basin Initiative."

6. Idaho Conservation League, "Columbia Basin Initiative."

7. David Cannamela (fish biologist), in discussion with the author October 13, 2022.

8. Idaho Conservation League, "Columbia Basin Initiative."

9. Investopedia Team, "What Is the CARES Act," Investopedia, updated July 25, 2023, https://www.investopedia.com/coronavirus-aid-relief-and-economic-security-cares -act-4800707#:~:text=The%20CARES%20Act%20can%20be,and%20state%20and %20local%20government; Department of Defense, "Overview," USASPENDING.gov, September 29, 2022, https://www.usaspending.gov/agency/department-of-defense?fy=2022; Andrew Glass, "Bush Bails Out U.S. Automakers, Dec. 19, 2008," Politico, December 19, 2018, https://www.politico.com/story/2018/12/19/bush-bails-out-us-automakers-dec-19 -2008-1066932; Tara O'Neill Hayes and Katerina Kerska, "Primer: Agricultural Subsidies and Their Influence on the Composition of U.S. Food Supply and Consumption," American Action Forum, November 3, 2021, https://www.americanactionforum.org/research /primer-agriculture-subsidies-and-their-influence-on-the-composition-of-u-s-food -supply-and-consumption/.

10. Wendell Berry, "The Unforeseen Wilderness," *Hudson Review* 23, no. 4 (Winter 1970–71): 633–47, at 643.

INDEX

Alves, Molly, 128–29; public relations efforts, 118–19; representative of Tulalip Tribe, 124–25

Anacostia River: Alger Park, 38–40; Watts Branch, 253–54

Anderson, Nate: Cadey Creek project, 140–43; South Gilbert Creek project, 134–38

Atlantic White Cedar bogs: Arden Bog, 57–59; description of, 45–46; Howard's Branch restoration, 47–51

bankfull discharge: development of, 16–17; identification of, 13, 29–30

basket weaving and making: Eastern Band of Cherokee Indians' use of, 163; Morgan, Dylan, 174–78; sourcing River Cane, 164, 172, 175

beaver dam analogues (BDAs), 122

beaver impacts: conflicts with homeowners, 106–7; Gilbert Creek, 139; historical perception of, 119–20; relation to stormwater control, 183; stream channel evolution, 122–23; Sumner Creek, 148–49; upon trout, 132

beaver trapping. See Collins, Dylan

brook trout: reserves, 135; stream restoration approach for, 136, 142–43

Cain, Rodger. See River Cane, historical accounts of

Chattahoochee Brick Company, 207–8, 212

Chesapeake Bay: Chesapeake Bay Program (EPA), 56; restoration, 43

climate change: impacts from drought, 152–53; impacts to springs, 158; impacts to stormwater control, 151; impacts to Wisconsin trout streams, 131; Lick Run, 99; Milwaukee Metropolitan Sewerage District goals, 198; modifying stream restoration techniques, 135; role of Atlantic White Cedar forests, 46

Collins, Dylan: beaver trapping process, 108–10

combined sewer overflows: Lick Run, 99–100; South River, 220–22

Cop City (Atlanta), 218–19

Eastern Cherokee Band of Indians. See basket weaving and making and River Cane, historical accounts of

Eisenhauer, Bill, 204–6

Elofson, Mel, 234–38

Elwha River Ecosystem and Fisheries Restoration Act, 232, 243

Elwha River restoration: access road, 248–50; role of sediment, 227–28, 237, 252; salmon recovery, 233, 236, 241, 243

Faifax, Emily, 120–21

Griffith, Adam, 160–62; role with Eastern Cherokee Band of Indians, 163–66, 175, 177

Guitierrez, Esperanza, 190–94

Hawley, Bob: interaction with Rosgen, 97; log structure installation, 91–95; role in Lick Run daylighting, 100–101

Hogland, Bruce. See River Cane, historical accounts of

Johnson, Kent: monitoring with Trout Unlimited, 145–47; Pine Creek monitoring, 155–56, 158–59; Sumner Creek monitoring, 149–51; work with River Falls, 147–48. *See also* Trout Unlimited

Kinnickinic River (Milwaukee), 190–93
Kinnickinic River (River Falls): monitoring of, 144–47; MS4 permit, 147–48
Kondolf, Matthias, RiverLab training, 36

Lave, Rebecca, study of Rosgen system/trainings, 30–34
Lewis, Logan, 71, 77, 83
Lick Run: costs, 98, 101; design considerations, 99–101; history of area, 99
Lincoln Creek (Milwaukee), 187–88
Lititz Run, 67–68

McHenry, Mike, 230–34
Merritts, Dorothy: awards and career, 70; teaching style, 75–76
Mitro, Matt, 130–32
Moore, Annie, 208–11

Noibi, Yomi, 199, 212–13

permitting and easements: costs in Wisconsin, 141; Elwha River, 232; for log jam structures, 94–95; Nate Anderson (Wisconsin Department of Natural Resources), 138; for regenerative stormwater conveyance, 55–56; use of River Cane, 167–69
Peters, Roger, 239–44
Pine Creek: brown vs. brook trout, 155–56; design considerations, 154; groundwater contamination, 158; monitoring results, 155
Proctor Creek, 198–202

regenerative stormwater conveyance: bubbler, 62–63; constructability, 52; expanding use of, 60; iron flocculate, 54–55; mitigating stormwater impacts, 52–53, 65–66; Saint Luke's, 60–62
Rhanis, Mike, 71, 83–85
river cane: flowering, 165; growth, 162–64, 166; propagation of, 169–70
river cane, historic accounts of: Cain, Rodger, 172; at Cowpen's Historic Battlefield, 170;

on Eastern Cherokee Band of Indians' lands, 165; Hogland, Bruce, 173–74; Roosevelt, Teddy, 161

Rosgen, Dave: classification system, 19–21; criticism of, 25–27, 36; role in permitting, 25; teaching style, 22–24; trainings, 14–15, 24–25

salmon recovery. *See* Elwha River restoration
sediment, legacy: accounting for sediment and nutrient reductions, 81–82, 86; Big Spring Run, 80–83; Bruebaker Run, 85–88; criticism related to, 78–79; discovery, 72–73; plant communities and seedbank, 77, 82; prioritization, 75, 77, 83–85; relation to stormwater, 79–80, 86–87
Shafer, Kevin: background, 185–86; management style, 196–97
South River: consent decrees concerning, 221–22; designated uses, 224–25
South River Forest, 216–20
Stephens, Donna, 207–8
stream restoration costs: Elwha River, 232; Pine Creek, 154–55; Proctor Creek, 203; Wisconsin Department of Natural Resources, 137
stream restoration stability/warranty: impacts from beavers, 116–18; legacy sediment projects, 77; Wisconsin Department of Natural Resources, 135

Taylor Creek mitigation site, 11–113; "Rebarn Again" program, 113
Trout Unlimited: chapter meeting, 130–33; partnership with Nate Anderson, 138; work on Pine Creek, 157
Tucker, Al, 211–12

Underwood Creek (Milwaukee), 181–84
Underwood, Keith: business property, 40–42; construction techniques, 51–53; controversy surrounding, 54–57; permitting of projects, 42–44, 51, 55, 59

Vanderhoof, Jennifer, 112–15, 126–27; planning for beaver manual, 111, 113–14

Walter, Bob, 70
Wohl, Ellen, 96; research on beaver dams and fire, 121

Printed and bound by CPI Group (UK) Ltd, Croydon, CR0 4YY

22/05/2024

14505621-0005